The
Apple
Book

The Apple Book

ROSANNE SANDERS

Philosophical Library · New York

PUBLISHED IN ASSOCIATION WITH THE ROYAL HORTICULTURAL SOCIETY

To H.B.

Published in the United States of America in 1988
    by Philosophical Library, Inc.
200 West 57th Street, New York, N.Y. 10019

First published 1988
Phaidon Press Limited, Oxford, 1988
Text and illustrations © Rosanne Sanders 1988
Apple Growing © Harry Baker 1988

Library of Congress Cataloging-in-Publication Data
Sanders, Rosanne.

  The apple book
  Includes Index.
  1. Apple–England.   2. Apple–Varieties.
  3. Apple–Pictorial works.   I. Title.
SB363.2.G7S26 1988   634'.11'0942   88-17875

ISBN 0-8022-2559-4

Printed in Great Britain by
The Roundwood Press Limited
Kineton, Warwickshire

ACKNOWLEDGEMENTS

During the early stages of the preparation of this
book, I received enormous help from Duncan Met-
ford of Bicton College of Agriculture and I would
like to express my thanks to him for his interest and
the assistance that he gave me in providing some of
the samples for the illustrations. I am enormously
grateful as well to the orchard staff at Wisley who
allowed me to plunder their orchard for the great
majority of specimens that I painted and to the
librarian there, Judith Jepton, for her help in the
library. Warm thanks to the following people who
also supplied me with fruit and blossom: Mr Con-
nabeer of Dartington, Mr Fairweather and George
Ball of East Charleton, Tom Griffiths of Dartington,
Wilson Gough of Dittisham, Dick Fulcher of Bicton
College of Agriculture and the Slapton Ley Fruit
Farm. My thanks to Dr Robb-Smith for his informa-
tion on the Blenheim Orange. Thanks also to my
friends, particularly Henry Will, Alan and Peter,
who supplied invaluable assistance in all sorts of
other ways. Finally my deepest gratitude to Harry
Baker without whose invaluable advice and help this
book would never have been completed.

BIBLIOGRAPHY

Baker, H., *The Fruit Garden Displayed*, 1986, Royal
    Horticultural Society, Cassell
Beach, S.A., *The Apples of New York. Vol 1 & 2*,
    1905, Albany, New York
Bultitude, John, *Apples*, 1983, Macmillan Reference
    Books
Bunyard, Edward A., *A Handbook of Hardy Fruits.
    Apples and Pears*, 1920, London
Hills, Lawrence D., *The Good Fruit Guide*, 1984,
    Henry Doubleday Research Assoc.
Hogg, Robert, *The Fruit Manual, 5th Edition*, 1884,
    London
Maund, B., *The Fruitist*, 1845–51, Groombridge &
    Sons, London
Roach, F.A., *Cultivated Fruits of Britain*, 1985, Basil
    Blackwell. U.S.A., Oxford
Scott, J., *The Orchardist*, c. 1873, London
Simmons, A.F., *Simmons' Manual of Fruit*, 1978,
    David & Charles
Smith, Muriel W.G., *National Apple Register of the
    United Kingdom*, 1971, Ministry of Agriculture,
    Fisheries & Food. London
Taylor, H. V., *The Apples of England*, 1948, Crosby,
    Lockwood, London
Gardeners Chronical, 1883, 1884, 1886, 1900,
    1908, 1930
The Herefordshire Pomona, 1878–85
*Transactions, Journals & Proceedings of the Royal
    Horticultural Society*, 1805 onwards

CONTENTS

PUBLISHER'S PREFACE

What a pleasing thing an apple tree is. Whether in breathtaking spring bloom or heavily laden with fruit, the apple tree, perhaps more than any other growing thing, conveys to us a sense of peace and well-being, bounty and blessing. How many youngsters, or adults for that matter, have found refuge from care or a haven for dreams in its welcoming branches? How many of us have been entranced by the sight of apple blossoms in moonlight after a spring rain, their scent drifting on the breeze? An apple a day may not be enough to keep the doctor away, but biting into a crisp, juicy apple is as refreshing for the spirit as the body.

How long has the apple tree been with us? Wild apples like the crab have probably been around since fruit trees first began to grace the planet. Experts have yet to agree, however, on when and where the cultivated apple first made its appearance, southern Europe and western Asia being the most likely points of origin. What is certain is that whatever the date and location of its birth the cultivated apple has been with us as a delightful part of the human diet for thousands of years.

The love affair with the apple most likely began with the Romans, who discovered the art of grafting and budding fruit trees and cultivated many new varieties from the eastern half of their vast empire. The results of these fruitful experiments were spread throughout Europe as the Roman forces increased the boundaries of the Empire ever northward until it reached its farthest outpost—England. In medieval times fruit culture became firmly established and by the thirteenth century several apple varieties had come into existence. Then English apple culture continued rather haphazardly until the start of the 1800s, when Thomas Andrew Knight pioneered a scientific approach to apple breeding.

The cultivated apple probably reached our shores along with the early settlers. One of the most devoted of apple lovers became one of our most beloved folk heroes: The American pioneer John Chapman, better known as Johnny Appleseed. A far gentler devotee of apple growing than the Roman legions of the Caesars, Chapman gave or sold seedlings and trees to families who were leaving his native Pennsylvania to participate in the great western migration. Eventually he traveled to Ohio, planting apple seeds along the way. Until his death in 1845 Johnny Appleseed traveled back and forth along his original route, visiting woodland orchards and tending his trees, teaching those who were among the creators of the hundreds of apple varieties thriving in America today. (The United States does not recommend the use of systemic insecticides for food crops.)

Everyone is familiar with the ubiquitous Red and Golden Delicious apple varieties and the McIntosh, available in every supermarket across the United States. Other popular, if not so widely available varieties, are the Cortland, raised in 1898 by S.A. Beach at the New York State Agricultural Station in Geneva, New York; the Rome Beauty which originated with H.N. Gillett in 1848, Lawrence County, Ohio; the Newton Pippin of Gershom Moore, which was popular in 1759 on Long Island, and the Northern Spy. Many old varieties, though not available commercially, are available through nurseries. Among other important American apples are the Rhode Island Greening, the York, the Winesap varieties, the Red Astrachan, the Macoun, the Roxbury Russet, the Lady and the Esopus Spitzenberg.

The Apple Book is a beautiful and unusual tribute to one of the sweeter aspects of life. We hope you will find as much pleasure in reading it as we have had in bringing it to you.

INTRODUCTION

We inherit the history of several thousand kinds of apple that have been considered worthy of a name, yet today the average Englishman knows perhaps ten. This has not always been the case. In Victorian times fruit was proudly offered at the dinner table for epicures to savour and discuss. It is this sad reflection on the passiveness of the present that prompted me to begin this book, hoping to encourage others to take a greater interest in the delicious fruits that can be found with a little judicious hunting. Even the smallest garden has room for a couple of apple cultivars, and the chapter on growing apples by Harry Baker will help those who wish to start a fruit garden.

I became fascinated by apples through painting them. No other fruit has the variety of colour, form, pattern or texture. At first I looked only at these external features and soon realized how little I knew about them. So I began studying apples, gradually learning what to look for and how to understand their characters.

It is an aim of this book to expand the reader's knowledge of the apple. There will, I hope, be readers who already have an understanding and who will wish to share with me the visual splendour of the apple. There will be those for whom the whole subject is a complete mystery and to whom I hope this book will prove a guide and inspiration.

I trust that this book will help those planning an orchard to decide which cultivars they want and how to grow them; it should also help those who have inherited apple trees to discover what it is that they possess and how to care for them. It is a celebration of my love for the apple, of its beauty and of its beneficence, which I hope you will share, so that together we may recapture some of the lost romance.

Rosanne Sanders

INDEX OF APPLES

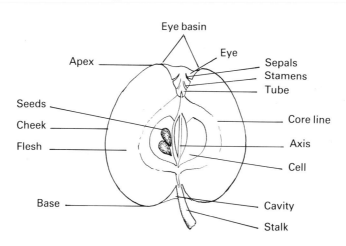

THE IDENTIFICATION OF APPLES

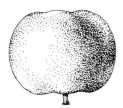

Flat-round

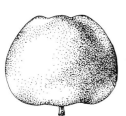

Round-conical

Long-conical

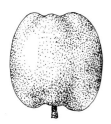

Oblong

Season

The apples are listed according to season, starting with the earliest. The season given is that during which the fruit can be eaten at its best. The early cultivars will not keep and cannot be stored, the mid-season cultivars will only keep for a short period, but the late cultivars only mature some time after they are picked and need to be stored before they reach their full flavour.

Ripening of the fruit is subject to regional and seasonal variation, so precise dates for picking are not possible. The best test for readiness is to lift the fruit gently and give it a slight twist. If it parts easily from the spur, leaving the stalk intact, it is ready to be picked.

Fruit Characteristics

Size

The size was taken from the fruit of old mature trees. Fruit taken from young trees are almost invariably larger than that quoted, especially if they are on dwarfing rootstocks. An average size has been estimated and given as diameter first and height second. The sizes are grouped into seven categories.

Fruit size will vary according to the climate, soil and size of the crop, and the sizes given are intended only as a general guide.

King Fruit

This fruit lies at the centre of the truss. Being the central flower it is the first to open and the fruit is often larger than the others. It sometimes has a fleshy protuberance at the stalk end and often the stalk is very short and thick. When picking the fruit for identification it is better, if possible, to avoid the king fruit as it may be misleading.

Shape

The shapes have been defined under eight headings. The shape of an apple can be a fairly regular guide, being only slightly influenced by climate, soil or crop size. Fruits from the northern latitudes are sometimes slightly more conical. When examining the shape of an apple, it should be held with the stalk (base) downwards and the eye (apex) uppermost.

Ribs are apparent on some cultivars. Up to five ribs can run between base and apex.

Five crowned Sometimes the ribs become a pronounced 'crown' at the apex.

Symmetrical When the sides are equally developed the fruit is symmetrical.

Lop-sided When the sides have developed unequally.

Regular When a horizontal section appears to be nearly circular.

Irregular When a horizontal section is angular, elliptical or irregular.

Waisted When the curve is concave towards the apex, for example Laxton's Early Crimson.

Skin

The description is intended to define a good example of the fruit showing the characteristics normally found on that particular cultivar. The examples chosen are fruits that have been well exposed to the sun and not those which have been shaded, as these will tend to be greener and lacking in the characteristic colours. Inevitably variances will be found according to the amount of sun and the type of soil. The examples illustrated show ripe fruits as they appear on the tree. Some fruits become yellower when stored and some become more greasy and shiny, and if so this is stated in the text. The numbers in brackets indicate a reference to the Royal Horticultural Society's colour chart.

Flush This is an area of almost unbroken colour which can appear as a small patch or almost wholly cover the fruit.

Mottling When the overlying colour is broken showing the skin beneath.

Stripes The fruit can be striped with varying shades of red. These stripes can be well defined long unbroken stripes, such as appear in Jupiter, or short broken stripes or splashes, as in Queen.

Scarf skin This term is used to describe a thin whitish layer of skin giving a rather milky appearance to part of the fruit, usually around the base, as in Grenadier and Jupiter.

Hammering Sometimes the surface is slightly pitted, rather like beaten copper, and feels slightly uneven. This is a characteristic sometimes observed in cultivars such as Jonathan and Annie Elizabeth.

Flat

Round

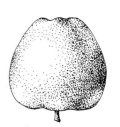

Conical

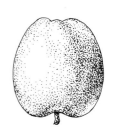

Oblong-conical

Hair line Some cultivars occasionally have fruits that bear a narrow sometimes russetted hair line that runs between base and apex. It is not a malformation but a characteristic of that cultivar. It is found, for example, in Keswick Codlin.

Russet This is a brown, somewhat cork-like layer that forms over the skin of the apple. Russet can appear in small or large patches, small scuffs, or resembling scattered dust. It can almost entirely cover the surface of the fruit and in this case the cultivar is known as a Russet apple, for example Egremont Russet. The amount and location of russet on the fruit is important for identification; for example, on Rosemary Russet it is mainly concentrated around the apex.

Reinette A group of apples with colour, and with areas of russet but not in sufficient quantity for the apple to be classified as a Russet apple. The term was originally intended to denote a fruit of quality, as indicated by the finest Reinette of all, the Cox's Orange Pippin.

Netting A network of russet can sometimes extend over part of the surface of the fruit, as in Lord Hindlip.

Lenticels are pores that are irregularly distributed over the surface of the fruit through which gasses interchange between the interior and exterior surfaces. They are an important feature of identification. They are usually roundish but can be angular or star-shaped, as is sometimes found in Sunset and Charles Ross.

Areolar In some cultivars, such as Granny Smith, a halo of colour can surround the lenticel.

Texture A fruit can be said to be smooth, as in Lane's Prince Albert, or rough as in Cornish Gilliflower. It can be dry as in Grenadier, or greasy as in Bramley's Seedling. This grease is a wax produced naturally by the fruit, as opposed to the protective wax used commercially, and often increases after the fruit has been picked and stored.

Bloom In some cultivars, such as McIntosh Red and Red Delicious, the surface of the skin is covered with a fine whitish bloom. This can easily be rubbed off so is more apparent when the fruit is still on the tree.

Stalk

The identifying points about the stalk are length and thickness. These can vary between fruits of any cultivar but there is a general expectation in most cases.

Sometimes however definite variations occur as in Ellison's Orange, where the stalk can either be fairly long and thin, or short and stoutish. Some cultivars have fruits with a fleshy bump on the stalk which is a characteristic and not a malformation, for example Newton Wonder. The stalk of the King fruit (see sub-heading 'Size') can be fleshy or swollen.

Cavity

The depression round the stalk is known as the cavity. The depth and width and the presence or absence of russet is important.

Lipped Some fruits within a cultivar can develop a lip on one side of the cavity which often results in the stalk being set at an angle. Good examples are Jupiter and Charles Ross.

Eye

This is the only part of the flower remaining after the fruit has been formed. After the petals drop the fruit develops and when mature the five sepals take up one of five distinct forms.

Erect convergent in which the sepals are standing upright with their margins touching and their points meeting together at a common point.

Erect in which the sepals are standing upright but not meeting together.

Connivent in which the sepals are standing upright with their points meeting together but overlapping each other.

Flat convergent in which the sepals lie flat with their points facing inwards and their margins touching.

Divergent in which the sepals are quite reflexed and fall back on to the basin.

The eye can be open, partly open or closed.

Reflexed tips The tips of the sepals can be bent outwards away from the eye.

eye

Basin

The depression in which the eye is situated is called the basin. Ribs may be visible within the basin and sometimes the skin is puckered around the eye.

Beading In some cultivars, such as Merton Worcester, up to five small round fleshy knobs are present in the basin round the eye.

Downy Varying amounts of white downy hairs can be present on the sepals.

Internal Characteristics

Robert Hogg in *The Fruit Manual* based his classification on the internal characteristics of the fruit and I have followed his principles of description: the shape of the tube, the position of the stamens within the tube, the shape of the carpels or seed cells and their relationship to the axis. All these internal characters, with the exception of the relationship of the cells to the axis, can be seen in the drawn cross section of each cultivar.

Tube The cavity just beneath the eye, (a) cone-shaped, (b) funnel-shaped.

Stamens (a) marginal – when they are situated at the outer margin of the tube, (b) median – when they are situated approximately in the middle of the tube, (c) basal – when they lie near the base of the tube.

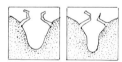

Tube

Stamens

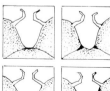

Core line

Axis

Seeds

Upright

Upright-Spreading

Spreading

Core line A line surrounding the core is usually clearly visible and its point of attachment to the tube can vary: (a) basal – meeting, (when the lines either side of the tube meet and touch at the base of the tube), (b) basal – clasping (when the lines either side of the tube join the base of the tube but do not touch), (c) median – (when the lines either side of the tube join the tube in the middle), (d) marginal – (when the lines either side of the tube join the tube at the point nearest the eye).

Core The position of the core can vary. It can be close to the stalk (sessile), at the centre of the fruit (median), or far from the stalk (distant).

Cells If the seed cell is split through the middle longitudinally it will usually be seen to be one of five shapes: (a) Round – (when the widest part is central and the cell roundish in shape), (b) Ovate – (when the widest part of the cell is nearest the base of the apple), (c) Obovate – (when the widest part of the cell is nearest the eye of the apple), (d) Elliptical – (when the widest part of the cell is central but the cell is elliptical instead of round), (e) Lanceolate – (when the cell tapers to a point at both ends).

Tufted In some cultivars the inside of the seed cells have some white woolly ribboning, for example Cornish Gilliflower and Lane's Prince Albert.

Axis If a horizontal section is cut, the attachment of the cells to the axis can be seen. If the cells are symmetrical they are axile, whether (a) open or (b) closed. If the cells are asymmetrical they are (c) Abaxile.

Seeds They can be (a) Acuminate – (long and sharply pointed), (b) Obtuse – (short and blunt), (c) Acute – (somewhere between the previous two).

Flesh Variations in the colour and texture of the flesh are usually constant within a cultivar. The taste of fruit is subjective and can again vary according to soil and climate.

Aroma This usually indicates the scent of the fruit before it is cut, unless otherwise stated.

Tree Characteristics

Flowers

The grouping numbers refer to the flowering time and are listed from 1–7 on the pollination table.

Triploid

Plants grow by the multiplication of cells brought about by cell division. When the cell divides, thread-like structures can be seen in the nucleus which are called chromosomes. The number of chromosomes found during cell division is characteristic of a species. The majority of apple cultivars grown today contain thirty-four chromosomes and are known as diploid. When the cell divides, the chromosomes separate into two equal sets of seventeen, one set donated by the male parent, the other by the female. In the case of a triploid the number of chromosomes is fifty-one and cannot be evenly divided and it is found that the larger set is usually donated by the female and that the pollen is sparse. Triploids are therefore not used as pollinators, so it is necessary to plant three cultivars within the same group, one to pollinate the triploid and one to pollinate the pollinator and triploid.

Leaves

There are two types of leaf, the leaves attached to the spur and the leaves of the new growth shoots. It is the former that is described. The leaf shapes are divided into six categories: (a) Oval (b) Broadly oval (c) Narrow oval (d) Acute (e) Broadly acute (f) Narrow acute.

The teeth along the leaf margins may be (a) Serrate – (toothed in a saw-like manner) (b) Biserrate – (when the serrations are again serrated) (c) Crenate – (with rounded convex teeth) (d) – Dentate – (with vertical rather then angled teeth).

The surface of the leaf may be flat or undulating and the sides can be upward-folding or downward-folding. The underside of the leaves are variably covered with pale downy hairs. With some cultivars the leaves tend to hang in a downward drooping fashion and that is referred to as downward-hanging.

Growth

Different cultivars vary in their habit of growth and if reasonably left to assume their natural shape will fall into roughly six categories. (a) Upright (b) Upright-spreading (c) Spreading (d) Wide-spreading (e) Compact (f) Weeping.

Cultivars also differ in their habit of bearing fruit and it is important to ascertain the habit before pruning.

Spur system Trees that produce their fruit buds on two year old wood and on spurs on the older wood. Spurs are short jointed lateral branches carrying fruit buds. Each flower bud contains four or five flowers that should produce fruit. Behind that bud a new bud will form and the growth will continue in that way until after some years a multiple spur system will have formed.

Tip-bearer A habit found in some cultivars of producing most of their fruit from the terminal fruit buds of shoots made the previous summer. These cultivars will produce few spurs on the older wood. A good example is Irish Peach.

Spur and tip bearer (partial tip-bearer) These cultivars will produce some fruit on the tips of shoots made the previous summer, as in tip-bearers, but they also produce fruit on spurs made on the older wood, as in a typical spur-bearer. A good example is Discovery.

Oval   **Broadly oval**

Narrow oval   **Acute**

Broadly acute   **Narrow acute**

Wide Spreading

Compact

Weeping

EMNETH EARLY

Season   Late July to mid August
Picking time   Late July and early August

An early Codlin-type cooking apple which was once widely planted commercially for the English markets. It was raised from Lord Grosvenor × Keswick Codlin by William Lynn of Emneth, Cambridgeshire, England and was introduced as Emneth Early by Messrs Cross of Wisbech. The first record of this apple was in 1899 when it received a First Class Certificate from the RHS. It later became known commercially as Early Victoria. It makes a hardy, compact tree of moderately weak vigour and upright habit. The cropping is heavy, though with a biennial tendency, and the fruits can be small unless severely thinned. It is a spur-bearer. The fruits have a superb flavour, sub-acid and sweet yet sharp when cooked. I found it did not need the addition of sugar and it cooks to a fluff. It is available from fruit tree nurserymen today, usually listed under Early Victoria.

Size Medium, though rather variable, 64 × 61mm (2½ × 2⅜″).
Shape Round-conical to oblong-conical. Can be slightly lop-sided. Irregular with prominent ribs.
Skin Yellowish-green (145A) becoming fairly bright greenish-yellow (5C). Frequently a hair-line present. Lenticels appear as small whitish-green dots which become larger towards the base and into cavity. Skin smooth and dry.
Stalk Medium thick (3mm) and quite long (23mm). Extends well beyond base.
Cavity Fairly wide and shallow.
Eye Small. Closed. Sepals quite large and long, erect and held tightly together with some tips slightly reflexed. Downy.
Basin Shallow and quite small. Well puckered with a trace of beading.
Tube Cone-shaped.
Stamens Marginal.
Core line Basal.
Core Median. Axile.
Cells Elliptical. Tufted.
Seeds Abundant. Acute. Plump and oval.
Flesh White with a slightly greenish tinge. Crisp, firm and juicy.
Aroma Almost nil.
Flowers Pollination group 3
Leaves Medium to fairly large. Oval to broadly oval. Crenate. Medium thick. Flattish and very slightly undulating. Light yellowish-green in colour with undersides quite downy.

GLADSTONE

Season   Late July to mid August
Picking time   Late July and early August

An attractive, very fragrant, very early dessert apple which is deliciously sweet and juicy when picked from the tree and immediately devoured, but should not be kept a moment longer otherwise it becomes soft and dry. It will not keep longer than a few days. It was introduced in 1868 by Mr Jackson of Blakedown Nursery, Kidderminster, Worcestershire, England, as Jackson's Seedling and was purported to be a chance seedling probably originating about 1780. In 1883 it received a First Class Certificate from the RHS, after which it was re-named Mr Gladstone, after the Prime Minister. The 'Mr' is now generally omitted. It makes a spreading tree of moderately vigorous growth which is a spur and tip bearer. It crops well but can be biennial.

Size Medium, 63 × 54mm (2½ × 2⅛″).
Shape Round-conical. Rounded at base. Large well rounded ribs with one sometimes larger. Irregular. Can be lop-sided. Can be five-crowned at apex.
Skin Variable in colour. Pale yellowish-green (144C to 150C). Partly to almost completely covered with deep red (45A) or brownish-red (179A) flush. Rather inconspicuous stripes of brownish-crimson (185A). Some fruits can have a much more streaked appearance without the dense flush. Occasionally a bold wide stripe of yellow skin colour can show through the flush. Lenticels conspicuous as grey-brown russet dots on flush or purple-brown dots on green skin. Skin smooth becoming greasy after picking.
Stalk Medium thick (3mm) and variable in length (9–16mm). Protrudes beyond base.
Cavity Fairly narrow and fairly shallow. Usually some golden-brown slightly scaly russet present which may come out on to the base.
Eye Medium size. Closed or slightly open. Sepals erect convergent.
Basin Fairly deep and rather narrow. Ribbed. Russet free. Slightly downy.
Tube Cone-shaped.
Stamens Median.
Core line Median towards basal.
Core Median. Axile, open.
Cells Ovate.
Seeds Rather small. Acute. Light or golden brown in colour. Very plump, rather rounded and pointed.
Flesh White tinged green. Soft. Coarse-textured.
Aroma Very fragrant.
Flowers Pollination group 4. Biennial.
Leaves Small to medium. Broadly oval. Serrate. Medium thick. Slightly upward-folding and undulating. Mid grey-green in colour with undersides very downy.

LAXTON'S EARLY CRIMSON

Season   Late July to mid August
Picking time   Late July and early August

This very early dessert apple was raised in England in 1908 by Laxton Brothers of Bedford from Worcester Pearmain × Gladstone. It was introduced in 1931. It is no longer grown commercially and may only be found in one or two specialist nurseries today. It makes a spreading tree of weak to moderate vigour and is a partial tip-bearer. The fruits are sweet and rather dry with a pleasant flavour. The cropping is moderate and the fruit should be eaten straight off the tree or within a few days of picking.

Size Medium, 67 × 63mm (2⅝ × 2½").
Shape Conical, sometimes long-conical. Rather waisted near apex. Large well-rounded ribs which end in a very well pronounced ridge around the eye. Can be slightly lop-sided but usually fairly symmetrical. Fairly regular.
Skin Dirty yellowish-green (between 146D and 144C) becoming dull pale greenish-yellow (150C). Three quarters to almost completely covered with deep crimson flush (53A) which is a paler crimson-red (46A) on the shaded side and is slightly striped and mottled over the yellow skin. There can be some short broken stripes of purplish-crimson (187B) on the paler flush, otherwise no stripes. There may be some scarf skin at base. Lenticels are tiny crimson dots but noticeable only on yellow skin. Occasionally a few small ochre russet surface patches.
Stalk Fairly slender to medium (2.5–3mm), sometimes thickening towards base. Fairly long (17–21mm). Protruding well beyond base.
Cavity Shallow and fairly narrow. Usually yellowish-green with some scaly golden-brown russet which can streak out over base.
Eye Medium size. Closed. Sepals connivent, fairly long and tapering, sometimes separated at base.
Basin Narrow and deep with well pronounced crown which can sometimes look like large beads. There may be some scaly ochre-brown russet present.
Tube Cone-shaped.
Stamens Median.
Core line Sometimes two, median and marginal.
Core Median to slightly sessile. Axile.
Cells Roundish to roundish obovate, occasionally roundish ovate.
Seeds Small, fairly plump. Acute.
Flesh Greenish-white. Soft. Fine-textured. Dryish.
Aroma Very sweetly scented.
Flowers Pollination group 2.
Leaves Medium to small. Narrow acute. Serrate. Thick and leathery. Very upward-folding, smooth not undulating. Mid green. Undersides downy.

GEORGE CAVE

Season   Early to mid August
Picking time   Early to mid August

An early dessert apple which is grown to some extent commercially today and can be found at farm shops. It was raised in 1923 in Essex, England by Mr George Cave at Dovercourt, and is said to be a chance seedling. It was acquired by Messrs Seabrook & Sons Ltd. of Boreham, Essex but was not named until 1945. It is widely available from fruit tree nurserymen today. The tree is fairly hardy, upright-spreading in habit and of moderate vigour. The tree is a spur-bearer. The fruits have no particular distinction: they are crisp, juicy and rather acid and the skin is a little tough. The cropping is good though the fruit quickly drops.

Size Medium small, 57 × 51mm (2¼ × 2").
Shape Round-conical. Trace of well-rounded ribs. Regular. Symmetrical or slightly lop-sided.
Skin Pale yellowish-green (150C) becoming yellow (8A). Half or more covered with rather sparse red flush (42B). Quite a lot of broken crimson-red stripes (46A) which can also appear on the green or yellow skin. There can sometimes be some scarf skin especially at base which can radiate from cavity as ragged dots and extend over flush giving it a rather mottled appearance. Lenticels conspicuous as large pale ochre russet dots. There can be some small greenish-ochre russet patches. Skin smooth and dry.
Stalk Medium thick (3mm). Quite long (15–20mm) extends well beyond base.
Cavity Rather shallow, medium width. Usually some scaly grey-brown russet which can extend over base. Often lipped.
Eye Medium size, partly open. Sepals broad based, long and slightly connivent, partly reflexed with some tips broken off. Downy.
Basin Shallow. Medium width. Some puckering.
Tube Cone-shaped.
Stamens Median.
Core line Median towards basal.
Core Median. Axile.
Cells Obovate.
Seeds Acuminate. Fairly plump and fairly regular.
Flesh Creamy-white, sometimes tinged slightly green under skin. Slightly soft. Fine-textured. Juicy.
Aroma Slightly sweetly aromatic.
Flowers Pollination group 2.
Leaves Medium size. Acute to rather narrow acute. Serrate. Medium thick, slightly leathery. Slightly upward-folding. Dark grey-green. Undersides very downy.

BEAUTY OF BATH

Season   Early August
Picking time   Early August

Beauty of Bath used to be one of the most important very early dessert apples grown commercially. The tree has a tendency to drop the fruit before it is fully ripe, therefore straw used to be placed beneath the 'Bath' trees in order to lessen the damage to the falling fruit. It was introduced in 1864 by Mr George Cooling of Bath, who states that it is a Juneating seedling originating about the mid 1800s near Bath. It received a first class certificate from the RHS in 1887. Today it is not grown commercially but is widely available from specialist fruit tree nurserymen. The tree is slow to come into bearing and can be irregular in cropping due to its early flowering. It makes a hardy, moderately vigorous, spreading tree that produces spurs freely. It is fairly resistant to scab. The fruits have a full flavour, sweet with a slightly acid tang.

Size Medium, 63 × 51mm (2½ × 2″).
Shape Flat-round. Regular. Sometimes a hint of ribs, usually noticeable at apex and occasionally at base. Symmetrical or lop-sided.
Skin Pale whitish-green (145C) to yellow (12B). Variably flushed and dotted with bright red (45A). Traces of short broken stripes of brownish-crimson (185A). Numerous large yellow lenticels give the flush a very mottled appearance. Usually some scarf skin at base which can extend over flush on cheeks. Some grey-brown russet streaks or patches. Skin smooth and greasy.
Stalk Stout (4mm) and short (9–12mm). Usually within cavity or protruding slightly beyond.
Cavity Fairly deep and wide. Scarf skin within cavity. Sometimes some fine grey-brown russet but often russet free.
Eye Can be almost closed with broad-based sepals erect and pressed together, even connivent, but often the tips are well reflexed leaving the eye half open and the stamens visible.
Basin Wide and fairly deep. Usually slightly ribbed. Sometimes some fine grey-brown russet. Downy.
Tube Funnel-shaped.
Stamens Median towards marginal.
Core line Median.
Core Median. Axile.
Cells Round.
Seeds Obtuse. Plumpish. Straight not curved.
Flesh Creamy-white, sometimes tinged pink near flush. Rather coarse textured. Juicy and soft.
Aroma Very sweetly aromatic and fruity.
Flowers Pollination group 2.
Leaves Medium size. Broadly oval. Crenate or bluntly serrate. Medium thick. Flat. Dark blue-green. Undersides not very downy.

IRISH PEACH

Season   Late August to early September
Picking time   Late August

This apple is supposed to be of Irish origin but from exactly where it came or who raised it remains a mystery. It was introduced into England in 1820 by Mr John Darby of Addiscombe and Mr Robertson of Kilkenny. It is an attractive, fine quality, second early dessert apple. If caught right it has a lovely rich vinous flavour with plenty of juice, but it quickly goes soft and dry. It is best when eaten straight from the tree and will not keep. The tree is of moderate vigour, spreading in habit and is a pure tip-bearer forming little or no spurs. It is a rather gaunt looking tree. The cropping is irregular. It is available today from several specialist fruit tree nurserymen.

Size Medium, 61 × 48mm (2⅜ × 1⅞″).
Shape Round rather flattish to slightly round-conical. Fairly conspicuous broad ribs which can sometimes be rather angular. Some fruits rather flat-sided. Can sometimes be slightly five-crowned at apex. Rather irregular. Usually fairly symmetrical.
Skin Pale yellow tinged green (150C) with very mottled brownish-orange flush (172C). Broken often broad stripes of milky-crimson (42B). Lenticels very conspicuous on yellow skin appearing as grey-brown or greenish-russet dots and small triangles: less conspicuous on flush where they appear more ochre-brown. Occasionally some patches of ochre-brown finely scaled russet. Skin smooth and rather greasy.
Stalk Rather stout (3–4mm) and sometimes fleshy. Medium length (15mm).
Cavity Medium to small and quite shallow. Usually some fine scaly brown russet radiating from stalk.
Eye Fairly small. Closed. Sepals connivent or erect convergent. Slightly downy.
Basin Wide. Medium depth. Ribbed and sometimes beaded.
Tube Cone-shaped.
Stamens Median.
Core line Almost basal.
Core Median. Axile slightly open.
Cells Broad ovate or obovate. Tufted. Rather transparent looking.
Seeds Roundish oval, acute. Straight not curved.
Flesh Creamy-white tinged very slightly green. Soft. Fairly juicy.
Aroma Strongly scented, sweet and fruity.
Flowers Pollination group 2.
Leaves Medium size, quite broadly oval. Rather bluntly serrate. Medium thick. Flat not undulating. Very slightly upward-folding. Mid blue-green. Undersides very downy.

LADY SUDELEY

Season   Mid August to early September
Picking time   Mid August

A very handsome second early dessert apple, originally named Jacob's Strawberry, which was raised in England at Petworth, Sussex in 1849 by a cottager named Jacob. The present name of Lady Sudeley is said to have been bestowed upon it by Lord Sudeley who, when having an extensive orchard laid out with this variety, was reminded of a striking dress worn by his wife at Court which somewhat resembled the bold striping and glowing colour of this apple. In 1884 the variety was given the Award of Merit by the RHS and was introduced in 1885 by George Bunyard & Co. It used to be grown to some extent commercially. It is worth growing for the beauty of its fruit alone and is available from a few specialist nurserymen. It makes a compact neat tree of moderate vigour. It is a tip-bearer, and the cropping is good. It is a useful early variety to grow in areas subject to late frosts because of its late flowering. The fruit has a good flavour when picked but quickly becomes dry and woolly with a slightly acid after-taste.

Size Medium, 67 × 57mm (2⅝ × 2¼″).
Shape Round-conical to oblong-conical. Flattened at base and often five-crowned at apex. Fairly distinct ribs. Can be lop-sided. Fairly regular.
Skin Pale greenish-yellow (154D) becoming golden yellow (8C). Half to almost completely covered with bright orangey-red flush (34A). Very prominent long broad stripes of bright crimson (46B) sometimes appearing over yellow skin. Occasionally some small russet patches. Lenticels very conspicuous as large greenish-brown or pale ochre russet dots. Can be some patchy scarf skin at base. Skin smooth and dry becoming greasy.
Stalk Medium thick (3mm) and short (5–10mm). Usually set within cavity.
Cavity Medium width and fairly deep. Often green and always russetted with fairly fine ochre or scaly brown russet which can streak over shoulder.
Eye Medium size closed or half open. Sepals erect convergent. Often stamens present.
Basin Deep and fairly narrow. Prominently ribbed. Sometimes a small amount of ochre-brown russet.
Tube Cone or funnel shaped.
Stamens Almost marginal.
Core line Median.
Core Median. Axile.
Cells Oval tending towards obovate.
Seeds Numerous. Fairly large, plump. Acuminate.
Flesh Creamy yellow. Soft and juicy.
Aroma Sweet and delicate, suggesting raspberries.
Flowers Pollination group 4.
Leaves Medium size. Acute. Serrate. Medium thick. Flat. Mid grey-green. Undersides very downy.

STARK'S EARLIEST

Season   Late July to mid August
Picking time   Early August

An early dessert apple of American origin, discovered at Orofino, Idaho in about 1938 by Douglas Bonner. It was assigned to Stark Bros. Nurseries & Orchards Co. of Louisiana, Missouri and introduced in 1944. The parentage is unknown. The fruit ripens at about the same time as Beauty of Bath but remains in good condition for longer. It crops heavily though the fruit is apt to be small and the early flowering makes it unsuitable for colder areas. It makes a moderately vigorous, upright-spreading tree that produces spurs freely. It is resistant to scab. This variety has lost popularity for commercial use, having been superseded by Discovery, but is still available from one or two specialist nurseries. It is frequently listed under the name of Scarlet Pimpernel. The cropping is good and the fruit is of quite good flavour.

Size Medium-small,   57 × 45mm   (2¼ × 1¾″)   or 60 × 60mm (2⅜ × 2⅜″).
Shape Variable. Flat-round to round-conical, sometimes conical. Slightly irregular. Usually slightly lop-sided. Hint of well-rounded ribs, sometimes showing at base otherwise fairly flattened at base.
Skin Very pale creamy-white (9D). Flush can appear as a light dusting of pinkish crimson (48A) or the fruit can be partly or almost completely flushed with bright crimson (53B), fading at the edges to milky pink. Lenticels very distinctive on flush as large pale ochre dots with a deep crimson surround and on the ground colour large russet dots with a slightly greenish or pink surround. Skin smooth and dry.
Stalk Fairly slender (2.5mm). Medium length (18mm). Protrudes beyond base.
Cavity Medium width and depth. Streaked with greenish-ochre and fine golden-brown scaly russet which usually streaks out over base.
Eye Fairly large. Sepals broad, erect convergent, rather pushed together with tips well reflexed.
Basin Rather narrow. Medium to shallow. Puckered and slightly downy.
Tube Rather long funnel-shaped.
Stamens Marginal.
Core line Median towards marginal.
Core Median. Axile open or abaxile.
Cells Round. Can be rather triangular.
Seeds Plump. Obtuse. Slightly curved.
Flesh White. Fine-textured. Fairly juicy. Crisp but tender. Skin very tender.
Aroma Fairly strong and pleasantly aromatic.
Flowers Pollination group 1.
Leaves Medium to smallish. Acute. Serrate. Medium thick. Flat not undulating. Mid to rather lightish green. Undersides moderately downy.

KATY

Season   September to early October
Picking time   Early September

Correctly known as Katja, this second early dessert apple was raised in Sweden in 1947 by the Fruit Breeding Institute at Balsgård from James Grieve × Worcester Pearmain. It was selected in 1955, introduced in 1966 and named in 1968. It is quite widely grown commercially as it is fairly hardy, a good pollinator and more vigorous than Worcester Pearmain. It makes a upright-spreading tree which spurs freely. The setting is heavy and requires thinning in most years. This Apple does not taste quite as good as it looks. The flavour is fair, slightly acid and refreshing but the skin is rather tough. Trees are available from several fruit nurseries.

Size Medium, 66 × 60mm (2⅝ × 2⅜″).
Shape Conical. Base rather rounded, tapering evenly towards apex which can sometimes be slightly five-crowned. Can be slightly lop-sided. Some fruits very faintly ribbed. Regular.
Skin Very pale slightly greenish-yellow becoming very pale primrose yellow (1C). Three quarters to almost completely covered with bright crimson-red flush (46A). Rather indistinct short broken stripes of crimson (53A). Areas of silverish-white dusting appear on the flush, mainly at base and apex. Lenticels very indistinct as tiny whitish or pale yellow dots. Sometimes some small ochre russet dots or patches. Skin smooth and dry. Rather dull on the tree but polishes to a fine shine. Becomes greasy if stored.
Stalk Medium to stout (3–4mm). Long (21mm). Nearly always set at a slight angle.
Cavity Medium depth and width. Even. Lined with fine greenish-ochre russet overlaid with some very fine scaly grey-brown russet which sometimes streaks a little over base.
Eye Fairly small. Closed. Sepals rather narrow, erect convergent with tips sometimes slightly reflexed. Stamens sometimes visible.
Basin Rather narrow and fairly shallow. Often beaded with rather large beads. Rather pinched looking. Usually russet free. Not very downy.
Tube Cone-shaped.
Stamens Median.
Core line Basal, clasping.
Core Medium to sessile. Axile.
Cells Roundish ovate.
Seeds Acute. Plump, slightly curved.
Flesh White tinged green. Fine textured and juicy.
Aroma Nil.
Flowers Pollination group 3.
Leaves Medium size. Narrow acute. Rather blunt, broad serrate. Medium thick. Upward-folding, not undulating. Mid green. Moderately downy.

DISCOVERY

Season   Mid August to mid September
Picking time   Mid August

The second early dessert apple is becoming very popular commercially and can be found in most shops today. It was raised in England in 1949 by Mr Dummer of Langham in Essex from Worcester Pearmain and possibly Beauty of Bath. Jack Matthews of Matthews Fruit Trees Ltd. in Thurston, Bury St. Edmunds, Suffolk, bought grafts of the tree which he named Thurston August. In October 1962 the variety was re-named Discovery and it was introduced in 1963. Discovery makes a moderately vigorous, upright-spreading tree and shows a resistance to scab. It is a spur and tip bearer. It is fairly hardy and the blossom shows good tolerance to frost, making it suitable for growing in colder areas. It is rather slow to come into bearing and has a tendency to drop quantities of fruitlets from young trees, but once mature crops heavily. The fruit has a remarkably long shelf life for an early dessert. The flavour is quite good.

Size Medium-small, 60 × 48mm (2⅜ × 1⅞″).
Shape Flat to flat-round. Well flattened at base and apex. Regular. Hardly any trace of ribs. Usually symmetrical.
Skin Pale greenish-yellow (1C) becoming more creamy-yellow (5D). Half to three quarters covered with brilliant crimson-red flush (46A). Some very indistinct carmine stripes (60A). Some scarf skin at base. Lenticels very conspicuous yellow or pinkish russet dots on flush, or less conspicuous tiny grey dots on yellow skin. Skin smooth and dry becoming slightly greasy.
Stalk Stout (4mm) and short (10mm). Level with base or protrudes slightly beyond.
Cavity Fairly wide. Medium depth. Lined with fine ochre-green and some light brown scaly russet, which can streak out a little over base.
Eye Medium size. Closed. Sepals erect convergent with tips reflexed or broken off. Downy.
Basin Moderately shallow. Medium width. Slightly ribbed. There can be some small beads.
Tube Cone-shaped.
Stamens Median.
Core line Basal.
Core Median. Axile.
Cells Round.
Seeds Acuminate. Fairly plump. Fairly regular.
Flesh Slightly creamy-white, sometimes tinged pink at flushed edges. Fine-textured. Crisp and juicy.
Aroma Almost nil.
Flowers Pollination group 3.
Leaves Medium to small. Acute. Serrate. Fairly thick and leathery. Flat not undulating. Upward-folding. Mid green. Undersides downy.

GRENADIER

Season   August to October
Picking time   Mid August

Grenadier is the standard early cooking apple in the United Kingdom and is extensively grown commercially for the early markets. It was brought to notice by Mr George Bunyard of Maidstone in Kent and first recorded in 1862, but it is thought to have been cultivated for many years prior to that, although there is no record of its origin. The variety was introduced commercially in 1875 and it received a First Class Certificate from the RHS in 1883. It makes a spreading tree of only moderate vigour which spurs freely. It is hardy and resistant to scab and canker but susceptible to capsid bug. It is suitable for growing in the North. The fruits are tangy and juicy and rather honey-flavoured. They cook to a fluff and are good for baking and purees. The cropping is heavy. It is widely available from nurserymen.

Size Large, 83 × 63mm (3¼ × 2½″).
Shape Round-conical, rather flattish. Distinct well-rounded ribs. Sometimes flat-sided. Rather irregular. Slightly five-crowned at apex. Symmetrical or lop-sided.
Skin Fresh pale green (144D) becoming pale yellowish-green (144C). Some scarf skin mainly at base. Lenticels conspicuous appearing as numerous white dots at apex and green or brown russet dots on cheeks. Skin smooth and dry.
Stalk Stout (4mm) and short (10–12mm). Usually fleshy. Does not extend beyond base and is set deep within cavity.
Cavity Wide and deep. Usually regular in outline but can have a large rib extending into cavity making it uneven. Sometimes a trace of grey-brown scaly russet (177C) which can streak out over shoulder.
Eye Small. Closed. Sepals connivent with tips reflexed. Fairly downy.
Basin Narrow and fairly shallow. Much puckered and ribbed.
Tube Cone-shaped.
Stamens Median.
Core line Basal to median.
Core Median. Abaxile.
Cells Elliptical.
Seeds Rather small. Mid brown. Straight. Acute.
Flesh White tinged green. Firm, fine-textured and juicy.
Aroma Almost nil.
Flowers Pollination group 3.
Leaves Medium size. Fairly broadly acute. Rather small bluntly pointed serrations. Fairly thick and leathery. Slightly undulating. Mid green. Downy.

LAXTON'S EPICURE

Season   Late August to mid September
Picking time   Late August

A high quality second early dessert apple, grown to some extent commercially. It was raised in England in 1909 by Laxton Brothers Ltd. of Bedford from Wealthy × Cox's Orange Pippin. It was awarded the Bunyard Cup in 1929 and 1932 and received the Award of Merit from the RHS in 1931. It is widely available from fruit tree nurserymen today. It makes a hardy, upright-spreading tree of moderate vigour and is a spur-bearer. The trees are hardy and frost resistant making them suitable for planting in colder areas. The fruits are sweet and juicy, somewhat pear-flavoured, and refreshing, though the skin is a little tough. The tree tends to over-crop with small fruits.

Size Medium, 63 × 54mm (2½ × 2⅛″).
Shape Flat-round to round-conical. Regular in outline with very slight trace of ribbing. Symmetrical.
Skin Pale yellowish-green (145A). Quarter to three quarters flushed with dull brownish-red (178D to 172B). Short broken stripes of dull brownish-crimson (185A). Lenticels not very conspicuous appearing as small pinky-grey or pale ochre-grey russet dots. Skin smooth and dry.
Stalk Fairly slender (2.5mm), long (30–35mm). Extends well beyond base. Sometimes with a fleshy knob.
Cavity Wide and deep. Regular. Usually russet free but can have a small amount of fine pale grey-brown russet at the base of the cavity.
Eye Medium size, closed or partly open. Sepals broad-based and connivent with tips reflexed.
Basin Medium width and fairly shallow. Puckered and slightly ribbed.
Tube Cone-shaped.
Stamens Median.
Core line Median.
Core Median. Axile.
Cells Roundish or elliptical tending towards ovate.
Seeds Quite large and numerous. Mid brown. Acute, fairly long but bluntly pointed. Slightly angular.
Flesh Creamy-white, sometimes tinged pink. Rather coarse-textured. Slightly soft but firm, not woolly. Juicy.
Aroma Faint when uncut, vinous when cut.
Flowers Pollination group 3.
Leaves Medium size, broadly acute. Serrate to broadly serrate. Thick and leathery. Slightly upward-folding and slightly undulating. Mid to dark grey-green. Undersides not downy.

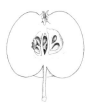

Season   Mid August to mid October
Picking time   Mid August

Season   Late August and early September
Picking time   Mid August

The origin of this old second early Codlin-type cooking apple is unknown. It is recorded as 'new' in *Scott's Orchardist*, a catalogue of fruits cultivated at Merriott in Somerset, *c.* 1873, and was once extensively grown in gardens and orchards. It was superseded by Early Victoria and is now found only in private collections. It makes an upright-spreading tree of rather weak vigour and is a spur-bearer. It is fairly hardy and suitable for growing in the colder areas. The fruits are acid with a fair flavour and they break up completely when cooked, though not to a fluff. Rather yellow when cooked. The cropping is very heavy.

Miller's Seedling is a fairly well known second-early dessert apple that was raised in England in 1848 by Mr James Miller, a nurseryman, at Newbury in Berkshire. It used to be widely grown for the markets and had a very high reputation in Berkshire. It is still popular with some growers for marketing at the end of August but has a very limited availability from specialist nurseries. It received the Award of Merit from the RHS in 1906. It makes a moderately vigorous, upright-spreading tree and is a spur-bearer. The cropping is prolific, though it tends to be biennial, and the fruits can be small unless well thinned. The fruits are easily bruised. They have a distinctive though not strong flavour, being sweetly tangy and slightly scented. They are pleasant and refreshing. A red sport exists which is said to mature a week later.

Size Medium-large, 73 × 64mm (2⅞ × 2½″).

Shape Round-conical to conical. Definitely ribbed. Irregular and rather angular in shape. Frequently lop-sided.

Skin Pale yellowish-green (145B) becoming yellow (12B). Occasionally there is a tinge of pinkish-brown, otherwise no flush or stripe. A hair line can be present. Some scarf skin at base. Lenticels fairly conspicuous whitish-green dots which become larger towards the base and frequently enter stem cavity. Occasionally some pinkish-brown russet dots within the white dots. Skin smooth and dry.

Stalk Medium thick (3mm) and long (25–28mm). Protrudes beyond base.

Cavity Wide and fairly deep. Scarf skin within cavity. Occasionally some grey-brown russet within. Usually slightly ribbed.

Eye Large. Closed. Sepals connivent and pressed well together with tips reflexed. Downy. ˙

Basin Small and shallow. Extremely pinched looking as though the skin has been gathered around the eye. Ribbed and beaded.

Tube Long, rather straight cone-shaped.

Stamens Marginal.

Core line Median.

Core Median. Abaxile.

Cells Elliptical or slightly ovate. (Hogg in *The Fruit Manual* found them Ovate).

Seeds Light golden-brown. Fairly plump. Acute.

Flesh White tinged green. Firm. Fine-textured, fairly juicy.

Aroma Nil.

Flowers Pollination group 3.

Leaves Medium size, broadly oval. Broadly, slightly bluntly serrate. Fairly thick and leathery. Slightly undulating. Mid green. Undersides slightly downy.

Size Medium, 63 × 54mm (2½ × 2⅛″).

Shape Round-conical. Slight trace of ribs and sometimes rather flat-sided. Fairly regular. Symmetrical or slightly lop-sided.

Skin Pale greenish-yellow (150C) becoming cream (5D). Flush can be a slight blush of pale orangey-pink (31C) or can be bolder and brighter mottled milky-pink (39A) with short broken bright, slightly milky red stripes (47A). Lenticels inconspicuous as whitish dots. Skin smooth and slightly greasy.

Stalk Slender (2mm) and long (18–26mm). Protrudes well beyond base of fruit.

Cavity Deep and wide. Regular. Can be russet free or can have some fine ochre-brown russet, sometimes overlaid with some fine grey-brown scaly russet streaking from the stalk.

Eye Small. Closed. Sepals quite large, broad based, connivent with tips reflexed. Very downy.

Basin Medium width, rather shallow sometimes with the eye almost standing on top of the fruit. Ribbed and puckered, sometimes beaded.

Tube Cone-shaped.

Stamens Marginal.

Core line Median.

Core Median. Axile wide open or abaxile.

Cells Elliptical to Ovate.

Seeds Rather flat and roundish. Straight acute.

Flesh White. Soft but crisp. Very juicy. Fairly fine-textured. There is a trace of pink in the flesh when fully ripe.

Aroma Very slight, pleasantly sweet.

Flowers Pollination group 3. Inclined to be biennial.

Leaves Medium to small. Broadly oval. Serrate. Medium thick. Flat and very slightly undulating. Mid yellowish-green. Undersides downy.

DUCHESS'S FAVOURITE

Season   Late August and September
Picking time   Late August

A very pretty second early dessert apple, raised in England by Mr Cree, a nurseryman of Addlestone in Surrey, somewhere between the late 1700s and early 1800s. It was named in response to the acclaim with which it was received by the then Duchess of York. It used to be grown in the Kentish orchards for the London markets. Trees are available from one or two specialist nurseries. The trees are moderately vigorous, upright-spreading, spur-bearers, and very fertile. The fruits are sweet with a slight 'bite', possessing a pleasant though not strong flavour. The skin is rather tough and chewy.

Size Medium-small, 58 × 51mm (2¼ × 2″).
Shape Flat-round. Flattened at base and apex. Very occasional trace of ribs. Regular. Symmetrical.
Skin Pale greenish-yellow (154C) becoming pale yellow (5D). Half to almost completely covered with bright red flush varying from light red (42A) or a light orangey-red (34A) on shaded side, to crimson-red (46A) on sunny side. Some faint short broken stripes more noticeable on the paler flush. Large pale ochre lenticels becoming more numerous, smaller and whitish towards apex. Skin shiny and smooth, becoming greasy.
Stalk Slender (2mm) and fairly short (12mm). Level with base or protruding slightly beyond.
Cavity Medium width and moderately deep. Lined with fine golden russet which streaks out over base. There can be some scarf skin which makes pinky-white streaks or mottling over the flush around the base.
Eye Small to medium. Open. Sepals quite long if not broken off. Sepals erect with tips well reflexed. Some stamens usually visible. Downy.
Basin Shallow. Wide. Puckered and slightly ribbed.
Tube Mostly cone-shaped. (Robert Hogg in *The Fruit Manual* found the tube to be funnel shaped).
Stamens Median towards marginal.
Core line Median. Sometimes red.
Core Median. Axile.
Cells Round, slightly ovate or slightly obovate.
Seeds Acuminate to acute. Numerous. Rather angular and slightly curved. Fairly plump.
Flesh White tinged red between skin and core line. Firm but tender. Slightly coarse-textured.
Aroma Nil. Very faint rather acid aroma when cut.
Flowers Pollination group 3.
Leaves Medium to small. Acute. Bluntly and broadly serrate. Medium thick. Slightly upward-folding, sometimes slightly undulating. Mid grey-green. Undersides not very downy.

DEVONSHIRE QUARRENDEN

Season   Late August to early September
Picking time   Late August

The origin of this very old second early dessert apple is uncertain. According to Bunyard in *A Handbook of Hardy Fruits*, this apple is mentioned by Worlidge in his *Vinetum Britannicum* in 1678, and probably takes its name from Carentan, an apple district in France. It is also mentioned in *The Compleat Planter and Cyderist* published in 1690. Some say it is a native of Devon where it was certainly a favourite and extensively cultivated. It thrives in the West Country as it will tolerate rain and wind and is still quite widely grown there. The tree is upright-spreading, moderatly vigourous and a spur bearer. It can be rather susceptible to scab. The cropping is moderate to heavy. The attractive fruits, have a lovely fruity flavour, sweet, juicy and crisp. Trees are available from a few nurseries.

Size Small, 51 × 38mm (2 × 1½″) to medium, 64 × 48mm (2½ × 1⅞″).
Shape Flat-round. Rounded at base and apex. Some ribs present which can be rather angular. Irregular. Symmetrical or lop-sided.
Skin Yellowish-green (145A). Three quarters to completely flushed with crimson-brown (178B) to dark brownish-purple (183B). Definite patches of green skin can remain where the skin has been shaded by another fruit or leaf. Lenticels inconspicuous pinky-grey dots on flush, conspicuous purple dots on green skin. Some rather indistinct brownish-crimson stripes (185A) noticeable at flush edges. Skin smooth and greasy.
Stalk Medium thick (3mm) and medium length (15mm). Protrudes beyond base.
Cavity Fairly wide. Medium to fairly shallow. Regular. Lined with some fine pale grey-brown russet.
Eye Large. Closed. Stamens long, connivent with tips well reflexed. Very downy.
Basin Wide and shallow. Sometimes the eye is almost sitting on top on the fruit. Puckered and lumpy, sometimes with irregularly shaped beads.
Tube Long funnel-shaped, extending up to core.
Stamens Marginal.
Core line Median.
Core Median. Axile open, occasionally. abaxile.
Cells Ovate.
Seeds Obtuse. Numerous. Regular.
Flesh Greenish-white, core-line and flesh sometimes red. Crisp and juicy. Slightly coarse.
Aroma Slight, pleasantly sweet.
Flowers Pollination group 2. Can be biennial.
Leaves Medium size. Acute. Sharply or bluntly serrate. Medium thick. Upward-folding but not undulating. Light grey-green. Very downy.

TYDEMAN'S EARLY WORCESTER

Season   Late August to mid September
Picking time   Mid August

This McIntosh type apple was raised in England in 1929 at East Malling Research Station in Maidstone, Kent, by H.M. Tydeman from McIntosh × Worcester Pearmain. It was introduced in 1945 and named Tydeman's Early in 1963. It makes a useful second early dessert apple for marketing just before Worcester Pearmain and it is grown commercially in several countries. Trees are moderately vigorous, very spreading with some long arching lateral branches. It bears most of the fruit on spurs but has a tendency to produce a small percentage of fruit on tips. The young tree will produce less fruit if the laterals are too severely pruned during the early years, otherwise the cropping is good. It is widely available from fruit tree nurserymen. The fruits have a good rich flavour, sweet with a slightly acid bite.

Size Medium, 67 × 60mm (2⅝ × 2⅜″).
Shape Round to round-conical. There can be a trace of ribs and sometimes one prominent one, otherwise fairly regular. Can be lop-sided.
Skin Greenish yellow (145A) becoming pale yellow (1C). Half to almost entirely covered with crimson red (46A) flush. Rather indistinct purplish-crimson (187B) stripes. On the shaded side it is frequently mottled and streaked with a paler greyed red (179A) with the yellow skin showing through. Lenticels numerous and reasonably conspicuous as pale pinky-white dots on flush and very pale yellowish-white dots on shaded side. Skin smooth and slightly greasy becoming more so if stored.
Stalk Fairly slender (2.5mm) and fairly long (17–20mm). Extends beyond base.
Cavity Fairly deep and rather narrow. Olive green lined with fine grey-brown very finely scaled russet which streaks a little over base.
Eye Small. Tightly closed. Sepals short, convergent with tips reflexed. Downy.
Basin Medium depth and rather narrow. Often beaded. Very slightly puckered.
Tube Cone-shaped.
Stamens Median.
Core line Median.
Core Median. Axile open, or abaxile.
Cells Obovate.
Seeds Acuminate. Straight. Fairly plump.
Flesh White. Fine-textured. Juicy.
Aroma Quite strong. Very sweetly scented.
Flowers Pollination group 3.
Leaves Medium size. Acute. Serrate. Medium thick and leathery. Not undulating but slightly upward-folding. Mid grey-green. Undersides downy.

KESWICK CODLIN

Season   Late September to October.
Picking time   Late August.

This very old second early cooking apple was found, according to Robert Hogg in *The Fruit Manual*, 'growing among a quantity of rubbish behind a wall at Gleaston Castle near Ulverstone' in Lancashire. It was introduced by John Sander, a Keswick nurseryman, in about 1790, and he named it Keswick Codlin. It is a typical codlin-type apple, being long and rather angular. The trees are prolific bearers, though with a biennial tendency. They are moderately vigorous, upright-spreading, spur-bearers and sufficiently hardy to be grown right up into the north of the country. They are available from a few specialist nurseries today.

Size Medium large, 74 × 67mm (2⅞ × 2⅝″).
Shape Round-conical to oblong-conical. Flattened at base tapering to a flattened apex. Prominent sometimes angular ribs often making the fruit rather flat-sided. Ribs become more pronounced towards apex where they terminate in five distinct crowns. Regular. Symmetrical or a little lop-sided.
Skin Pale green (145A) becoming pale yellow (150B). Can be slightly flushed with pale greyish-orange (167C) sometimes deepening to a greyish-brown (173C–174B). Flush often streaked in places. Some have one or more raised russet hair lines running from stalk almost to the apex. Lenticels prominent grey russet dots with pinky-brown surround on flush, otherwise green surround. Skin smooth and dry becoming slightly greasy.
Stalk Stout (4mm) and short (9mm). Often fleshly. Protrudes very slightly beyond base or level with base. Stalk sometimes incorporated within a large fleshy protuberance on side of cavity.
Cavity Wide and shallow. Regular. Some fine grey-brown russet radiates from stalk.
Eye Medium size. Closed. Rather pinched-in looking. Sepals broad based, long and connivent, with tips reflexed. Downy.
Basin Medium depth and width. Irregular. Often beaded. Some streaky brown russet can radiate out from eye.
Tube Cone-shaped.
Stamens Median.
Core line Basal.
Core Median to rather distant. Abaxile.
Cells Ovate. Lanceolate. Tufted.
Seeds Acute. Very plump. Tufted.
Flesh Yellowish-white. Soft. Rather coarse-textured.
Aroma Faintly acid.
Flowers Pollination group 2. Biennial.
Leaves Small. Acute. Serrate. Medium thick and leathery. Upward-folding and slightly undulating. Mid grey-green. Undersides downy.

REVEREND W. WILKS

Season   September and October.
Picking time   Early September.

A large, mid-season cooking apple which was raised in England from Peasgood Nonsuch × Ribston Pippin by Messrs Veitch of Chelsea. It was first recorded in 1904 when it received the Award of Merit from the RHS and it was introduced to commerce in 1908. It received a First Class Certificate in 1910. The Rev. W. Wilks was the Vicar of Shirley in Surrey and Secretary of the Royal Horticultural Society from 1888–1919. He was devoted to horticulture and probably the greatest gift that he gave to gardeners was the Shirley Poppy, *Papaver rhoeas*, which was raised by him. This variety makes a dwarfish tree of compact form, eminently suitable for small gardens. It has a good resistance to disease and is suitable for growing in the west. The trees are spur-bearers and produce spurs freely and the cropping is heavy with large fruits, though with a biennial tendency. The fruits are sub-acid with a delicate aromatic flavour and cook to a pale yellow froth. It is widely available from specialist nurseries.

Size Very large, 89 × 76mm (3½ × 3″).
Shape Round-conical to conical. Broad, flattened base tapering to a flattened apex. Symmetrical. Slightly ribbed and fairly regular.
Skin Pale whitish-green (145C) becoming pale primrose yellow (2C). Quarter to half slightly flushed with pale ochre (163C) or mottled pinky-red (35B). There can be some rather broken scarf skin at base. A few definite broad broken stripes of fairly bright milky-red (47A). Lenticels inconspicuous sparse grey-brown russet dots. Skin smooth, becoming greasy if stored.
Stalk Fairly slender (2.5mm). Short to medium (10–16mm). Within cavity or slightly beyond.
Cavity Wide and fairly deep. Slightly russetted with grey-brown sometimes slightly scaly russet.
Eye Medium size. Closed or partly open. Sepals erect or convergent. Stamens sometimes visible. Fairly downy.
Basin Deep and medium wide. Distinctly ribbed.
Tube Slightly funnel-shaped.
Stamens Median.
Core line Median.
Core Median. Abaxile.
Cells Roundish to elliptical.
Seeds Acute. Fairly plump, rather small.
Flesh White, crisp, fine-textured. Fairly juicy.
Aroma Nil.
Flowers Pollination group 2. Biennial.
Leaves Large to medium. Broadly oval. Serrate. Medium thick. Slightly upward-folding and slightly undulating. Dark green. Undersides slightly downy.

OWEN THOMAS

Season   Late August to early September
Picking time   Mid to late August.

This second early dessert apple was raised in England in 1897 by Laxton Brothers of Bedford. It was introduced in 1920. It makes a wide-spreading, moderately vigorous tree which is a spur-bearer. The cropping is light to moderate. It has a very short season: for a brief moment the fruits are quite nice and aromatic if eaten straight from the tree but in two or three days it is over. Trees have a very limited availability from specialist fruit tree nurserymen.

Size Medium-small, 57 × 51mm (2¼ × 2″).
Shape Flat-round. Rather rounded at base. Some large well-pronounced ribs making the fruit rather irregular in shape. Can be slightly five-crowned at apex.
Skin Yellowish-green (145A) becoming yellow (5B). Quarter to three quarters flushed with orange-red (168B) which can be rather sparse or can be in rather broad bands. Skin beneath flush or at the etremity is golden (163B). Some broken scarlet stripes (42A). Lenticels indistinct as small grey-green dots on green or yellow skin, pale pinkish-white dots on flush. There can be some fine grey-brown russet patches and some patchy scarf skin at base and on cheeks. Skin smooth and very slightly greasy.
Stalk Fairly stout (3.5mm). Short to medium length (7–15mm). Extends beyond base of fruit.
Cavity Medium to rather shallow depth and narrow. Sometimes lipped. There can be a small amount of fine ochre-brown russet radiating from the stalk which can continue slightly over base.
Eye Fairly small. Tightly closed or slightly open. Sepals small, erect or convergent with tips reflexed. Sometimes some stamens visible. Very downy.
Basin Medium width and medium depth. Much ribbed and puckered. There can be some beading.
Tube Funnel shaped.
Stamens Marginal.
Core line Median.
Core Median to slightly distant. Axile.
Cells Elliptical tending towards ovate.
Seeds Acute. Fairly plump, bluntly pointed. Straight.
Flesh Greenish-white. Fine textured. Some green transparent parts in the flesh. Soft. Fairly juicy.
Aroma Very slight, rather acid yet sweet.
Flowers Pollination group 2.
Leaves Small. Acute. Deeply cut serrate. Thick and leathery. Upward-folding and undulating. Mid blue-green. Undersides quite downy.

HARVEY

This is one of the oldest English culinary apples. It was mentioned as early as 1629 by John Parkinson, an apothecary who published details of the important fruit varieties available at that time. It was raised in Norfolk and thought to have been named after Dr Gabriel Harvey, Master of Trinity Hall, Cambridge. This variety was well known in East Anglia up to the middle of the present century. However, today it is no longer grown commercially, and trees have a very limited availability from nurseries. The tree is of moderate vigour, upright-spreading in habit and is a partial tip-bearer. The fruits are sub-acid with a good flavour and break up completely on cooking.

Size Large, 76 × 68mm (3 × 2⅝").

Shape Oblong-conical. Base flat, shoulders rounded. Flattened at apex. Broad well rounded ribs becoming more angular towards apex. Rather flat-sided. Can be irregular. Symmetrical or lop-sided.

Skin Yellow-green (144B) becoming yellow (7A). There can be a slight ochre-brown (152D) or greyed-orange (167C) flush. Frequently some patchy and spotted scarf skin at base and on cheeks. Lenticels fairly conspicuous pinky-brown or green dots surrounded by scarf skin becoming smaller and more numerous at apex. Small scuffs and patches of grey-brown russet on cheeks. Variably patched and netted with scaly grey-brown russet at apex and base, usually more on one side than the other. Skin slightly textured and dry.

Stalk Short or medium length (12–18mm). Medium thick (3mm). Level with or just beyond base.

Cavity Narrow and deep. Lined with scaly grey-brown russet which streaks and scatters over base.

Eye Medium to large. Open. Sepals short, convergent or erect convergent, sometimes slightly separated at base, with tips reflexed or broken off.

Basin Medium width and depth. Slightly ribbed. Some grey-brown slightly scaly russet. Slightly downy. Can be beaded.

Tube Funnel shaped.

Stamens Median.

Core line Median to basal clasping.

Core Median. Abaxile or axile open.

Cells Obovate.

Seeds Acute or acuminate. Broad and fairly plump.

Flesh Creamy-white. Firm. Fairly fine-textured. Dry.

Aroma Very slight, rather acid.

Flowers Pollination group 4.

Leaves Medium to small. Acute. Bluntly serrate or crenate. Medium thick. Flattish. Some slightly upward-folding. Light grey-green. Fairly downy.

GEORGE NEAL

This high quality second early culinary apple was raised in England in 1904 by Mrs Reeves at Otford in Kent. It was introduced in 1923 by R. Neal & Sons of Wandsworth in London and in that year received an Award of Merit from the RHS. The high colour of this apple gives the impression that it is not of high quality, hence it is not grown commercially which is a pity because the fruits have an excellent flavour, sweet yet a little acid; they cook pale yellow and remain in intact slices. It makes a tree of moderate vigour, spreading in habit and is a spur-bearer. The cropping is good. Trees have a limited availability from specialist fruit tree nurseries.

Size Large, 82 × 63mm (3¼ × 2½").

Shape Flat-round. Broad at base and apex. Slight broad well-rounded ribs. Frequently slightly lop-sided. Fairly regular.

Skin Pale green (144C) becoming pale greenish-yellow (2C). Orangey-brown (171A) with short broken stripes and dots of bright red. There can be some patches of greenish-grey russet. Lenticels fairly conspicuous as ochre-grey dots on flush and greyed brown russet dots or green dots on ground colour. Skin smooth and slightly greasy.

Stalk Fairly slender (2.5mm) and short to medium length (10–20mm). Within cavity or protrudes slightly beyond.

Cavity Narrow and deep. Lined with grey-brown scaly russet which can run concentrically round cavity and can come out over shoulder.

Eye Fairly large. Open. Sepals short and erect with some tips slightly reflexed. Slightly downy.

Basin Wide and deep. Slightly ribbed. Can be some tiny specks of grey-brown russet.

Tube Cone-shaped or slightly funnel-shaped.

Stamens Median.

Core line Median towards basal.

Core Median, slightly distant. Abaxile.

Cells Roundish.

Seeds Acute. Fairly plump. Straight.

Flesh Creamy-white. Rather soft. Fine-textured. Juicy.

Aroma Slight, rather acid.

Flowers Pollination group 2.

Leaves Medium size. Acute. Bluntly serrate. Medium thick. Very slightly upward-folding. Flat not undulating. Mid green. Undersides downy.

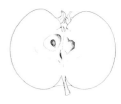

WORCESTER PEARMAIN

Season   September and October
Picking time   Early to mid September

This is a well known second early to mid season dessert apple of English origin which, although somewhat on the decline, is still important commercially. It was raised by Mr Hale of Swan Pool near Worcester and thought to be a seedling from a Devonshire Quarrenden. It was introduced in 1874 by Messrs Smith of Worcester and in 1875 received a First Class Certificate from the RHS. It makes an upright-spreading, round-headed tree of moderate vigour and is inclined to tip-bearing, producing few spurs. It is hardy and reliable and is a heavy and regular cropper. It shows a resistance to mildew but is prone to scab. It is suitable for growing in the colder areas. The fruits have a sweet and pleasant flavour which improves if the apples are left to ripen on the tree. Commercially they are picked too early resulting in inferior quality. Trees are widely available from nurserymen today.

Size Medium, 64 × 61mm (2½ × 2⅜″).
Shape Conical or round-conical, occasionally long-conical. Symmetrical or occasionally lop-sided. Slight hint of ribs. Regular.
Skin Pale greenish-yellow (between 1C and 145B) becoming pale yellow (3C). Almost completely covered with brilliant red flush (46B) with indistinct long stripes of crimson-red (46A). On shaded side; flush is pale red (43B) or pinkish red (39B) in long bands over yellow skin, with some brighter red stripes (45A). Lenticels distinct as ochre russet dots. Skin smooth and greasy.
Stalk Fairly slender (2.5mm) to fairly stout (3.5mm). Short to medium length (13mm).
Cavity Rather narrow and deep with stalk deeply inserted. Lined with fine greenish-ochre russet which can streak or scatter over base.
Eye Fairly small. Closed. Sepals erect, connivent with tips reflexed. Fairly downy.
Basin Shallow and narrow. Ribbed. Often beaded.
Tube Long funnel-shaped.
Stamens Median to marginal.
Core line Median.
Core Median. Axile.
Cells Obovate, sometimes slightly ovate.
Seeds Quite large. Acuminate. Fairly plump.
Flesh White. Rather coarse-textured. Fairly juicy. Firm and crisp but tender.
Aroma Strong and very sweetly aromatic.
Flowers Pollination group 3. Fairly frost resistant.
Leaves Medium size. Acute. Broadly serrate. Medium thick. Slightly upward-folding, not undulating. Mid green. Undersides downy.

MERTON KNAVE

Season   September
Picking time   Mid September

This very attractive and brightly coloured mid season dessert apple was raised in England in 1948 at the John Innes Institute by M.B. Crane. It was developed from Laxton's Early Crimson × Laxton's Epicure. It was originally named Merton Ace in 1968 and re-named Merton Knave in 1970. It makes a moderately vigorous, round-headed tree which is spreading and rather weeping in habit. It is a partial tip-bearer. The tree is hardy and the cropping is good although the fruit is not of particularly high quality. The flavour is pleasant but not strong, sweet with a nice acid tang. Trees are available from a few specialist nurseries today.

Size Medium-small, 58 × 48mm (2¼ × 1⅞″)
Shape Round. Sometimes slightly flattened at base and apex. There can be some indistinct well rounded fairly broad ribs but frequently no ribs. Regular. Usually fairly symmetrical.
Skin Greenish-yellow (151D). Three quarters to almost completely covered with brilliant red flush (46B) with some areas of crimson-red (46A). Indistinct broken stripes of carmine (53A) which are more noticeable on the shaded side. Lenticels fairly conspicuous small crimson or brown dots on shaded side, blackish purple dots on flush. There can by some very small ochre russet patches. Skin smooth becoming shiny and greasy if stored.
Stalk Oval and rather flattened. When viewed from the side it appears quite stout (3.5mm) but from the front, more slender (2.5mm). Length medium to very long (12–30mm). Extends beyond base.
Cavity Wide. Medium depth. Some fine greenish-ochre russet radiating out in thin streaks.
Eye Medium size, slightly open. Sepals connivent with tips well reflexed. Some stamens visible. Downy.
Basin Medium width and depth. Sometimes beaded. Slightly ribbed and puckered.
Tube Cone-shaped.
Stamens Median to marginal.
Core line Basal.
Core Median. Axile.
Cells Round.
Seeds Numerous. Angular. Acute. Fairly plump.
Flesh Creamy-white. Slightly coarse-textured. Can be tinged pink beneath the skin. Slightly tender. Fairly juicy.
Aroma Strong and fruity.
Flowers Pollination group 3.
Leaves Medium to small. Narrowly acute. Bluntly serrate. Medium thick. Very upward-folding. Slightly undulating. Dark grey-green. Downy.

JAMES GRIEVE

Season   September and October
Picking time   Early September

A very popular second early dessert apple, raised in Scotland by Mr James Grieve of Edinburgh, open-pollinated from Pott's Seedling or from Cox's Orange Pippin. It was introduced by Dickson's Nurserymen, employers of Mr Grieve and first recorded in 1893. It received the Award of Merit from the RHS in 1897 and a First Class Certificate in 1906. Several coloured sports exist. James Grieve is not extensively planted because the fruits bruise easily and may drop prematurely in warm districts. It prefers the North, disliking the humid West where it is prone to canker, otherwise it is hardy and adaptable. It makes a spreading, round-headed tree of moderate vigour. The flavour is excellent; sweet with a nice acid balance and fruits are acceptable as cookers in July. The skin is a little chewy. The fruit will keep until Christmas as a rather soft apple but with its flavour retained. Trees are widely available from nurserymen.

Size Medium, 70 × 60mm (2¾ × 2⅜″). Often very large on young trees.
Shape Round-conical to conical. Symmetrical or a little lop-sided. Very slight trace of ribs. Slightly irregular. Slightly rounded or flattened at base.
Skin Bright yellow-green (150B) becoming yellow (12B). Variably speckled and striped with bright red (46B) over orange skin (26A). Some small greenish-ochre russet patches. Lenticels fairly conspicuous as grey-brown russet dots. Skin smooth and greasy.
Stalk Fairly slender (2.5mm) to medium (3mm). Long (26–33mm). Protrudes beyond base.
Cavity Deep and wide. Greenish with some scaly grey-brown russet streaking out.
Eye Medium size. Closed. Sepals long, convergent and erect with tips reflexed. Sepals distinctly green with tips turning brown. Fairly downy.
Basin Medium width to rather narrow. Medium depth. Slightly puckered with slight trace of ribs. Some russet lying concentrically round basin.
Tube Funnel-shaped
Stamens Median.
Core line Median towards basal.
Core Median. Axile.
Cells Obovate. Sometimes roundish.
Seeds Quite large. Acute or acuminate. Fairly plump.
Flesh Creamy-white. Fine-textured. Soft and juicy.
Aroma Slightly sweetly scented.
Flowers Pollination group 3.
Leaves Medium size. Acute. Crenate or bluntly serrate. Medium thick. Sometimes slightly upward-folding. Dark green. Undersides downy.

MERTON WORCESTER

Season   September and October
Picking time   Early September

A second early dessert from the John Innes Institute in England, raised by M.B. Crane in 1914 from Cox's Orange Pippin × and Worcester Pearmain. It was named in 1947 and received the Award of Merit from the RHS in 1950. At one time it was fairly important commercially but is no longer planted due to its susceptibility to bitter pit. Trees are available from one or two specialist nurseries. The trees are reasonably hardy, moderately vigorous and upright-spreading in habit. They produce spurs very freely and the cropping is good and regular. The fruits have a good aromatic flavour, tasting slightly of pears and are sweet and juicy.

Size Medium-small, 58 × 54mm (2¼ × 2⅛″).
Shape Round-conical. Sometimes a slight trace of ribs and sometimes a little flat-sided. Fairly regular. Usually fairly symmetrical but can be a little lop-sided.
Skin Yellowish-green (145A) becoming greenish-yellow (154B). Half to three quarters flushed with brownish-red (179A) which can be ochre-brown (166C) towards the edges and on less well coloured fruits. Rather indistinct stripes of crimson-red (46A). Lenticels inconspicuous as minute russet dots. Some small patches of pale grey-ochre russet. Skin smooth and dry.
Stalk Medium (3mm). Medium to long (12-30mm). Extends well beyond base of fruit.
Cavity Medium width, medium depth with stalk deeply set. Usually some scaly light brown russet running round cavity.
Eye Fairly small, closed. Sepals finely tapered, convergent with tips reflexed or broken off. Fairly downy.
Basin Shallow. Medium width. Sometimes slighty ribbed. Usually five distinct beads.
Tube Small, funnel-shaped.
Stamens Median sometimes towards marginal.
Core line Basal.
Core Median. Axile, slightly open.
Cells Round.
Seeds Acuminate. Long rather narrow and pointed. Very dark brown. Starved looking not plump. Fairly straight.
Flesh Creamy-white. Firm and crisp. Fine-textured. Juicy.
Aroma Very slight, sweetly scented.
Flowers Pollination group 3.
Leaves Medium to small. Acute. Bluntly pointed and broadly serrate. Thick and leathery. Upward-folding and not undulating. Mid green. Undersides slightly downy.

MERTON CHARM

Season   September and October
Picking time   Mid September

A rather small but high quality mid season dessert apple. It was raised in England in 1933 at the John Innes Institute by M.B. Crane from McIntosh Red × Cox's Orange Pippin. It is an excellent garden apple but is too small for commercial use. It makes a spreading, slightly weeping tree of moderate vigour which produces spurs freely. It is a good regular cropper and is fairly hardy. The fruits have a good flavour, crisp, juicy and sweet. Trees are available from one or two specialist nurseries. It received an Award of Merit from the RHS in 1960.

Size Medium-small, 57 × 47mm (2¼ × 1⅞″).
Shape Round, rather flattened to round-conical. A slight trace of ribs. Flattened at base and apex. Sometimes lop-sided.
Skin Yellowish-green (145A) becoming golden yellow (151A). Quarter to three quarters flushed brownish-red (173A) to ochre-brown at extremities (164A). Overlaid with mottled and stripey brownish-crimson (185A) which in parts can fuse together to give quite a bright crimson-red flush. Some scarf skin at base. Lenticels fairly conspicuous especially on flush as tiny pinkish-white or yellowish-white russet dots. On green or yellow skin they are pinkish-brown russet dots. There can be some very small ochre-russet patches. Skin smooth and dry becoming slightly greasy.
Stalk Fairly slender (2.5mm) and shortish to medium length (10–15mm). Can be fleshy. Within cavity or protruding slightly beyond.
Cavity Quite narrow and fairly deep with stalk set well into it. Some very fine ochre-brown russet may radiate from stalk which often seems to occur in patches on opposite sides of the stalk.
Eye Small. Closed or partly open with some stamens visible. Sepals small and narrow, erect or convergent with tips reflexed. Downy.
Basin Narrow. Medium depth. Slightly ribbed.
Tube Tiny, cone-shaped.
Stamens Marginal.
Core line Basal. Clasping. Rather indistinct.
Core Median to very slightly distant. Axile.
Cells Roundish obovate.
Seeds Acuminate. Fairly large for size of apple. Mid brown. Not very plump. Slightly curved.
Flesh Creamy-white tinged slightly green around the edges. Crisp and juicy. Fine-textured.
Aroma Almost nil.
Flowers Pollination group 2.
Leaves Medium size. Narrow acute. Quite broadly serrate. Thin. Upward folding not undulating. Light yellowish-green. Undersides fairly downy.

SAINT EDMUND'S PIPPIN

Season   Late September and October
Picking time   Mid September

This early russet was raised in England about 1870 by R. Harvey at Bury St Edmunds in Suffolk, and is thought to be a chance seedling. It was first recorded in 1875, during which year it received a First Class Certificate from the RHS. It is the best early russet, though it has a limited commercial use as unfortunately the fruits bruise easily and can be unattractive where the russet is thin and patchy. The fruits have a first class flavour, very rich and quite sweet but with a nice balance of acidity. The cropping is uncertain and it can overcrop and bear small fruits. The fruits have a relatively short shelf-life. Trees are available from several specialist nurseries.

Size Medium, 63 × 52mm (2½ × 2″).
Shape Flat-round to round-conical. Well flattened at base and apex. Usually symmetrical but can sometimes be lop-sided. Sometimes a slight hint of one or two well-rounded ribs. Fairly regular.
Skin Pale greenish-yellow (1B) becoming golden-yellow (14B). Some fruits about a quarter flushed with fairly bright orange-red (171B). Partly to almost completely covered with fine greyish-golden russet (163C) overlaid with a fine scaling of brown. Lenticels inconspicuous as either tiny whitish dots, occasional dark brown dots or pale ochre dots on the russet which are often angular or star-shaped. Skin very dry and slightly rough.
Stalk Slender (2mm) and fairly long (12–22mm). Protrudes beyond base.
Cavity Deep and quite narrow. Lined with russet which circumvents the cavity.
Eye Fairly small. Closed. Sepals short, rather flat-convergent or connivent. Downy.
Basin Rather narrow. Medium depth. Frequently yellow or green skin showing just round the eye, otherwise russet lined. Slightly puckered.
Tube Cone-shaped.
Stamens Median or marginal.
Core Median. Axile.
Core line Median. Rather faint.
Cells Ovate.
Seeds Obtuse. Quite large. Blunt. Rather broad and frequently flattish on one side. Fairly plump and slightly curved.
Flesh Creamy-white. Tender and juicy. Fine-textured.
Aroma Very slight, sweetly aromatic.
Flowers Pollination group 2.
Leaves Medium size. Broadly oval. Bluntly serrate. Medium to thin. Can be slightly upward-folding. Not undulating. Light yellowish-green to mid green. Undersides slightly downy.

LAXTON'S FORTUNE

Season   September and October
Picking time   Early September

A mid season dessert apple raised in 1904 by Laxtons of Bedford from Cox's Orange Pippin × Wealthy. It was introduced in 1931, received an Award of Merit in 1932, and received a First Class Certificate from the RHS in 1948. It used to be grown commercially but is now considered too soft. It is a fairly small compact tree of moderate vigour, quite suitable for the small garden and is generally listed by fruit tree nurseries as Fortune. The trees are fairly hardy. They show some resistance to scab but are prone to canker, otherwise they are fairly trouble-free. The cropping varies with area: in some it crops regularly and colours well, but in others it tends to be biennial with poorly coloured fruits. It is suitable for growing in the North and West. The fruits are sweet with a good acid balance and good aromatic flavour.

Size Medium, 67 × 61mm (2⅝ × 2⅜″).
Shape Round, slightly conical. Slightly flattened at apex. Rounded at base. Symmetrical or slightly lop-sided. Some indistinct ribs more pronounced at apex. Five-crowned. Slightly irregular.
Skin Light green (144C) becoming yellow (2B). Quarter to almost completely covered with broken stripes, speckles and mottled areas of bright red (45B) with the yellow or ochre-yellow (162A) skin showing through. Some patches of ochre-brown russet. Lenticels inconspicuous tiny grey russets dots. Skin smooth and dry.
Stalk Fairly slender (2.5mm). Fairly long (25mm). Protrudes well beyond cavity.
Cavity Deep. Medium width. Sometimes lipped. Lined with greenish-ochre and some slightly scaly brown russet which streaks and scatters over base.
Eye Medium size. Closed. Sepals broad and long, connivent and tightly pinched together. Sepals are of a distinctive green and tend to remain green rather than turn brown. Not very downy.
Basin Narrow and deep. Ribbed. Sometimes beaded. Some cinnamon-brown russet may radiate out over apex.
Tube Cone-shaped, rather flattened at narrow end.
Stamens Median.
Core line Basal, clasping.
Core Median to slightly distant. Axile.
Cells Obovate.
Seeds Obtuse. Round, fairly plump. Straight.
Flesh Creamy-white. Tender but firm. Rather coarse-textured.
Aroma Very slight, sweetly scented.
Flowers Pollination group 3. Biennial.
Leaves Medium size. Oval to broadly oval. Small bluntly pointed serrate. Medium thick. Slightly undulating. Mid green. Not very downy.

MERTON BEAUTY

Season   September and early October
Picking time   Early September

Another second early to mid season dessert apple from the John Innes Institute of England. It was raised in 1932 by M.B. Crane from Ellison's Orange × Cox's Orange Pippin and released in 1962. It makes an upright-spreading tree of moderate vigour which produces spurs freely and it is a useful garden cultivar, being late flowering. The fruits are sweet and sharp with a distinct trace of aniseed, which it gets from Ellison's Orange. It is an extremely good apple and the cropping is good. Trees are available from one or two specialist fruit tree nurserymen.

Size Medium-small, 56 × 45mm (2³⁄₁₆ × 1¾″).
Shape Flat-round to very slightly conical. Usually no ribs, occasionally a slight trace. Regular. Usually symmetrical.
Skin Pale green (145B) becoming greenish-gold (151A). Quarter to three quarters flushed with mottled brownish-red (178B). Rather indistinct long broad stripes either of the same colour as flush or with a hint of carmine (178A). Small patches and larger rather fuzzy areas of light-brown or ochre-russet chiefly towards base but also on cheeks. Skin smooth and dry.
Stalk Medium thick (3mm) and long (20–27mm).
Cavity Medium width and depth. Lined with light brown scaly russet which orbits the cavity rather than radiates from stalk. It can straggle out over shoulder.
Eye Medium size. Slightly open. Sepals very long or broken off. Convergent at base with three quarters of the length well reflexed. Fairly downy.
Basin Fairly shallow, medium width. Slightly puckered.
Tube Cone-shaped, occasionally slightly funnel-shaped with a short funnel neck.
Stamens Basal to median.
Core line Basal.
Core Median. Axile.
Cells Obovate.
Seeds Rather large for size of fruit. Obtuse. Fairly broad, blunt ended. Not very plump, rather angular and slightly curved.
Flesh Creamy-white. Fine-textured. Juicy, crisp and firm.
Aroma Very slight.
Flowers Pollination group 5.
Leaves Medium size. Acute. Serrate. Thick and leathery. Slightly upward-folding, sometimes undulating. Mid green. Undersides not very downy.

ARTHUR TURNER

Season   September to November
Picking time   Late September to early October

This large early or mid season cooking apple was raised in England by Charles Turner of Slough, Bucks, and was first exhibited as Turner's Prolific at the RHS in 1912 where it was given an Award of Merit. It was re-named Arthur Turner in 1913 and introduced in 1915. The blossom is especially beautiful, and received the Award of Garden Merit from the RHS in 1945. The trees are vigorous, upright, are fairly resistant to scab, and are suitable for growing in the North. The trees produce spurs very freely and the cropping is regular and good. The fruits have a pleasant though not strong acidic flavour, and require a fairly long cooking time. They break up completely but not to a fluff. Trees are widely available from fruit tree nurseries.

Size Large, 80 × 48mm (3⅛ × 1⅞″) or 80 × 76mm (3⅛ × 3″).
Shape Round-conical to slightly oblong. Flattened at base and apex. Usually symmetrical. Slight, well-rounded ribs, more pronounced towards apex where it is rather five-crowned. Slightly irregular.
Skin Light green (145A) much overlaid at base with fine scarf skin. Variably flushed: from delicate blush of greyed orange (173D) to quarter flushed greyed red (179B). There can be a patch of bright purple pink at base (184C), and small flecks or dust-like specks of grey-brown russet. Lenticels are inconspicuous pinky-brown russet dots surrounded by white on the green skin. Skin smooth and dry.
Stalk Variable. Very stout (4–5mm) and short (10mm) deep within cavity, or medium thick (3mm) longer (16mm) protruding slightly beyond base.
Cavity Deep and usually wide. Lined with ochre-brown streaky russet, which can be slightly scaly.
Eye Medium size, partly open. Sepals can be slightly separated at base, rather flat convergent with tips reflexed. Quite downy.
Basin Quite deep and wide. Slightly puckered. There can be small amounts of brown russet.
Tube Long funnel-shaped.
Stamens Median towards marginal.
Core line Median.
Core Somewhat sessile. Axile.
Cells Round to slightly obovate. Frequently small.
Seeds Acute. Fairly wide. Fairly plump. Straight.
Flesh Yellowish-white. Coarse-textured, rather dry.
Aroma Nil.
Flowers Pollination group 3.
Leaves Large. Broadly oval. Bluntly serrate or crenate. Medium thick. Slightly upward-folding. Slightly undulating. Mid grey-green. Slightly downward-hanging. Undersides very downy.

ELLISON'S ORANGE

Season   September and October
Picking time   Mid September

This second early dessert apple was raised in England towards the end of the nineteenth century by the Rev. C.C. Ellison of Bracebridge in Lincolnshire and Mr Wipf, the gardener at Mr Ellison's brother-in-law's home, Hartsholme Hall, from Cox's Orange Pippin × Calville Blanc. On Mr Ellison's retirement, he offered to sell grafts of the tree for the financial benefit of Mr Wipf, and these were bought by Messrs Pennell & Sons. The variety was first recorded in 1904. It was introduced and received an Award of Merit from the RHS in 1911 and a First Class Certificate in 1917. It tends to crop biennially. It has a very rich and distinctive aniseed flavour and the cropping is good. The trees are hardy and are suitable for growing in the North, but it does not like areas of high rainfall. The trees are prone to canker but fairly resistant to scab. There are a number of red sports of this variety. It is widely available from fruit tree nurserymen.

Size Medium, 67 × 57mm (2⅝ × 2¼″).
Shape Round, slightly conical, sometimes slightly oblong. Flattened at base and apex. Usually symmetrical. Very slight trace of ribs. Regular.
Skin Dull yellowish-green (approx: 145B) becoming yellow (6B). Slightly to three quarters flushed with brownish-red (34B) with broad broken stripes of brownish-crimson (185A). Some small russet scuffs and patches. Lenticels inconspicuous tiny greenish-grey russet dots. Skin smooth and slightly greasy, becoming more so when stored.
Stalk Variable. Slender (2mm) and long (35mm) or medium-thick (3mm) and fairly short (15mm).
Cavity Medium to shallowish. Medium width. Lined with ochre and some fine light brown russet.
Eye Medium size. Tightly closed. Sepals erect convergent or slightly connivent with tips reflexed or broken off. Fairly downy.
Basin Medium width. Deep. Slightly ribbed.
Tube Rather small, cone or slightly funnel-shaped.
Stamens Median towards marginal.
Core line Basal clasping.
Core Median. Axile.
Cells Rather small. Roundish obovate or obovate.
Seeds Obtuse, rather broad, blunt and fairly plump.
Flesh Creamy white, tinged slightly green. Fairly fine-textured. Crisp and juicy.
Aroma Fairly strong. Rather vinous.
Flowers Pollination group 4.
Leaves Medium size. Acute. Bluntly and broadly serrate. Medium-thick. Slightly upward-folding, some slightly undulating. Mid green. Not very downy.

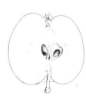

TOM PUTT

Season   September to November.
Picking time   Early September.

A very old variety of second early to mid season culinary apple which has always been very popular in the West Country where it was highly esteemed as a cider apple. It was raised in England towards the end of the 1700s by the Rev. Tom Putt, Rector of Trent in Somerset, who was reputed to be a keen fruit man. There appear to be considerable differences in growth between trees of the same name found in different counties and even in parts of the same county, probably due to the introduction of local seedlings taken from the original. The trees are very spreading, vigorous and produce spurs freely. The fruits are acidic, rather small for a culinary apple and liked by some as a dessert. The cropping is good and regular. Trees are available from some specialists today.

Size Medium, 65 × 55mm (2½ × 2⅛″).

Shape Flat-round to round-conical. Rather angular with well pronounced ribs especially noticeable at apex where they terminate in five often rather knobby crowns. Irregular. Sometimes symmetrical, sometimes lop-sided.

Skin Rather dull yellowish-green (144C) to ochre-yellow (153A). Half to three quarters covered with long sometimes unbroken broad stripes of slightly greyed-red (42B) to red (46B) which sometimes run together to produce an area of fairly intense rather stripey red (46B–46A). Lenticels very inconspicuous as tiny pinky-white or red dots on colour, whitish-green dots on green skin. Sometimes a hair line present. Skin smooth and almost dry becoming greasy if stored.

Stalk Stout (4mm) and rather short (7–15mm). Protrudes beyond base.

Cavity Fairly shallow to medium. Medium width. Usually some fine almost downy grey russet. Often a fleshy knob on one side. Lenticels often noticeable within cavity as white or pink dots.

Eye Quite large. Closed. Sepals connivent with tips reflexed or broken off. Downy.

Basin Medium width and depth. Well pronounced ribs. Very puckered and sometimes beaded.

Tube Cone-shaped.

Stamens Marginal.

Core line Median.

Core Median. Axile, wide open.

Cells Ovate.

Seeds Acute. Fairly large and quite plump.

Flesh Greenish-white. Firm, coarse-textured, crisp and juicy. Skin rather tough.

Aroma Sweetly scented.

Flowers Pollination group 3.

Leaves Large. Broadly acute. Serrate. Medium thick, undulating. Dark green. Downy.

AUTUMN PEARMAIN

Season   September to November
Picking time   Late September

This is one of the oldest dessert apples which has been in existence since the late 1500s. It is a second early to mid season apple of reasonable quality. Stocks of this cultivar are somewhat confused. It has been erroneously cultivated as the Royal Pearmain and has been found to be indistinguishable from the Herefordshire Pearmain. The trees appear to be no longer listed by nurserymen. The trees are moderately vigorous, upright-spreading and are partial tip-bearers. The cropping is moderately heavy and the fruits have a pleasantly sweet but not strong flavour.

Size Medium, 67 × 60mm (2⅝ × 2⅜″).

Shape Conical. Well rounded shoulders but slightly flattened base. Apex narrow and flattened. Symmetrical or a little lop-sided. Very indistinct well-rounded ribs, more pronounced towards apex. Slightly five crowned at apex. Fairly regular.

Skin Dull green (145A) changing to pale yellow (7D). Quarter to three quarters flushed dull gold (167A) to greyed-orange (168B). Broken stripes of red (45A) to dull reddish-brown (173A) away from the sun. Lenticels conspicuous large whitish russet dots. Much of the surface netted and flecked with fine greyish-ochre (approx. 162A) or grey-brown (199A) russet. Skin very slightly textured. Dry becoming greasy if stored.

Stalk Fairly stout (3.5mm). Medium length (12–17mm). Extends beyond base of fruit. Usually set at an angle due to lipped cavity.

Cavity Medium to rather shallow. Fairly narrow. Usually with large lip on one side. Partly or completely lined with fine brown or grey-brown russet which extends thinly over base.

Eye Fairly large, half or fully open. Sepals broad-based, erect convergent, sometimes flattish convergent with tips reflexed. Sepals usually separated at base. Very downy.

Basin Medium width. Fairly shallow. Some slight trace of ribs. Usually some russet.

Tube Wide cone or slightly funnel-shaped.

Stamens Median.

Core line Basal, clasping.

Core Median. Axile.

Cells Ovate (but obovate in Hogg's *Fruit Manual*).

Seeds Accuminate. Plump. Fairly straight.

Flesh Creamy-white tinged green. Fine-textured. Firm but tender. Fairly juicy.

Aroma Nil.

Flowers Pollination group 4.

Leaves Medium size. Broadly acute. Serrate. Medium thick. Slightly upward-folding, sometimes slightly undulating. Light rather yellowish-green. Undersides downy.

QUEEN ——————

Season   Mid September to December
Picking time   End August.

This is a fairly old mid season to late cooking apple.
It is of English origin and was raised in Billericay,
Essex, by a farmer named W. Bull. The apple was
raised in 1858, apparently from the pips of an apple
purchased in the market, and it first fruited in about
1874. The apple was introduced to commerce in
1880 as The Claimant by Messrs Saltmarsh of
Chelmsford and was awarded a First Class Certificate
from the RHS in November of that year. It is a
garden and exhibition cultivar and is still listed today
by one or two specialist nurseries, usually as The
Queen. The trees are moderately vigorous, upright-
spreading and partial tip-bearers. The cropping is
good and the fruits very acidic, tangy and fruity with
a very good flavour. It cooks to a fluff.

Size Very large, 89 × 64mm (3½ × 2½″).
Shape Flat. Distinctly flattened at base and apex.
   Well rounded ribs. Slightly irregular. Slightly
   five-crowned at apex. Fairly symmetrical, can be a
   little lop-sided.
Skin Very pale whitish-green (145A–145C) becoming pale whitish-yellow (2C). Quarter to three
   quarters flushed. The area of flush is made up of
   small flecks and dots of greyed-red (179A) rather
   than a dense flush and because the yellow skin
   shows through, the overall effect from a distance
   is a rather brownish-orange (172D). Very short
   broken stripes of bright red (45A). Lenticels
   inconspicuous as tiny green or grey dots. Skin
   smooth and very slightly greasy.
Stalk Medium thick (3mm). Fairly long (18mm).
   Usually protrudes beyond base.
Cavity Deep and very wide. Lined with fine pale
   ochre russet partly overlaid with scaly grey-brown
   russet which can streak out over base.
Eye Fairly large, half or completely open. Sepals
   erect convergent or somewhat divergent. Some
   stamens showing. Downy.
Basin Fairly deep. Medium width. Usually quite
   definitely ribbed.
Tube Cone-shaped.
Stamens Median.
Core Line Median.
Core Median to distant. Axile.
Cells Ovate or roundish.
Seeds Acute. Not very plump.
Flesh White. Fine-textured. Juicy and firm.
Aroma Very slight.
Flowers Pollination group 3.
Leaves Medium to large. Broadly oval. Bluntly
   pointed serrate. Medium thick. Flat or very
   slightly undulating. Mid green. Very downy.

WEALTHY ——————

Season   Mid September to December
Picking time   Mid September

An American mid to late season dessert apple, raised
by Peter M. Gideon of Excelsior, Minnesota, from
seed of the Cherry Crab which he obtained from
Albert Emerson at Bangor in about 1860. The
variety was imported into England during the 1800s
and was grown commercially for some while but is no
longer popular. In 1839 it received an Award of
Merit from the RHS. Trees are available from one or
two specialist nurseries. The trees are upright-
spreading of rather weak to moderate vigour, and
partial tip-bearers. The cropping is good and the
fruits are crisp, fairly juicy and fresh tasting with a
nice tang.

Size Medium,   67 × 61mm   (2⅝ × 2⅜″)   or
   67 × 54mm(2⅝ × 2⅛″).
Shape Round to flattish-round. Symmetrical. Some
   well-rounded ribs. Fairly regular.
Skin Very pale greenish-yellow (154C) becoming
   whitish-cream (5C). Quarter to nearly completely covered with pinkish-red flush (47A) which
   can be paler (47C) and rather striped. Many short
   wide broken stripes of crimson-red (46A) to
   crimson (53B) often extending into cavity.
   Whole appearance very striped except on well-
   coloured fruits. Lenticels fairly conspicuous as
   tiny pale ochre dots on flush, less conspicuous
   pale green dots on yellow skin. Skin smooth, dry
   and bloomed.
Stalk Very slender (1.5–2mm) and usually fairly long
   (18–22mm). Usually extends well beyond base.
Cavity Deep and narrow. Some slightly scaly ochre-
   brown russet.
Eye Small. Closed. Sepals connivent with tips
   reflexed. Not very downy.
Basin Deep and rather narrow. Ribbed and slightly
   puckered.
Tube Funnel-shaped or cone-shaped.
Stamens Median inclined towards marginal. Point of
   attachment above core line.
Core line Median, clasping neck of the funnel. Basal
   clasping on cone-shaped tube.
Core Distant. Axile.
Cells Ovate
Seeds Acute. Large and numerous. Long oval.
Flesh White tinged slightly green, sometimes tinged
   pink under skin. Tender. Rather coarse-textured.
Aroma Quite strong. Sweetly aromatic.
Flowers Pollination group 3.
Leaves Medium size. Acute. Serrate. Medium thick.
   Sometimes very slightly upward-folding and
   slightly undulating. Mid yellowish-green. Under-
   sides slightly downy.

LORD LAMBOURNE

Season   Late September to mid November
Picking time   Late September

This fairly well known mid season dessert apple is of English origin, having been raised by Messrs. Laxton Bros. of Bedford in 1907 from James Grieve × Worcester Pearmain. It was introduced by Laxtons in 1923. The RHS awarded it the Bunyard Cup in 1921 and an Award of Merit in 1925. It is grown on a small scale commercially in the UK. Trees are fairly widely available from specialist nurseries. The trees are compact in shape, of moderate vigour and partial tip-bearers. It is a heavy and regular cropper and it bears fruit of a uniform size. As it ripens it becomes greasy and tends to collect dust.

Size Medium 64 × 51mm (2½ × 2″).
Shape Round, slightly conical to flat-round. Usually symmetrical. There can be a slight trace of some well-rounded ribs. Regular.
Skin Pale greenish-yellow (4B) to yellow (8B), frequently with a layer of rather striped green scarf-skin over the top giving it a translucent appearance. Variably flushed: sometimes rather thin liverish-red (nearest being 172B with an overlay of green), but can be a fairly bright scarlet-red (42A). Short broken stripes of red (46A). Lenticels distinct as pale grey or green russet dots on flush, greenish-brown russet dots on skin. Skin smooth and slightly greasy, becoming more greasy if stored.
Stalk Fairly slender (2.5mm) to medium thick (3mm). Medium length (15–20mm). Protrudes beyond base.
Cavity Medium depth and fairly wide. Lined or partly lined with greenish-ochre and scaly brown russet which can streak out over base.
Eye Smallish, slightly open. Sepals convergent or slightly connivent, with tips reflexed. Very downy.
Basin Medium width, medium depth to rather shallow. Slightly ribbed, sometimes slightly beaded.
Tube Cone-shaped.
Stamens Median to marginal.
Core line Basal, clasping, sometimes meeting.
Core Median. Axile.
Cells Round.
Seeds Acute. Quite large and numerous. Rather broad.
Flesh Creamy-white, slightly coarse-textured. Firm but tender. Juicy.
Aroma Quite strong, sweetly scented.
Flowers Pollination group 2.
Leaves Medium size. Acute. Broad and bluntly serrate. Medium thick. Flat not undulating. Upward-folding. Dark grey-green. Undersides slightly downy.

HERRING'S PIPPIN

Season   Late September to early November
Picking time   Early September

This apple was raised in England by Mr W.A. Herring of Lincoln and first recorded in October 1908 when Mr Herring exhibited specimens at the RHS show. It was introduced to commerce by Messrs J.R. Pearson of Lowdham in Nottingham in 1917 and received the Award of Merit from the RHS in 1920. It used to be listed by most nursery firms but today is only available from a few specialist nurseries. It is a popular exhibition cultivar. The trees are moderately vigorous, upright-spreading and spurbearers. The cropping is good. It is an attractive dual-purpose apple with fair flavour, sub-acid with a hint of aniseed. It cooks to a dull yellow and stays reasonably intact.

Size Large, 83 × 76mm (3¼ × 3″).
Shape Round-conical to slightly oblong-conical. Symmetrical or slightly lop-sided. Large broad ribs, some of which are fairly well pronounced especially near apex. Five crowned at apex. Slightly irregular.
Skin Pale greenish-yellow (150C). Half to almost completely covered with red flush (45A) dense on sunny side, more mottled on shaded side. Rather indistinct broken stripes of deep slightly purplish-carmine (59A). Lenticels inconspicuous, rather sparse, small ochre or grey-brown russet dots. Much thin slightly streaky scarf skin especially at base and spreading partly up cheeks. Skin smooth, becoming very greasy when stored.
Stalk Stout (4–5mm). Very short (5–10mm). Sometimes fleshy. Sunk well into cavity.
Cavity Deep and fairly wide. Lined with fine greenish-ochre with a little fine brown scaly russet.
Eye Large, open, with fairly long erect sepals separated at base. Stamens visible. Slightly downy.
Basin Deep and usually quite wide. Ribbed. Regular.
Tube Cone shaped or slightly funnel shaped.
Stamens Median.
Core line Median to slightly basal. Frequently running almost up sides of the tube before branching out.
Core Median. Axile.
Cells Obovate.
Seeds Obtuse or acute. Fairly large and rather flat.
Flesh Creamy-white, sometimes tinged pink. Soft but fairly firm. Not very juicy. slightly coarse-textured.
Aroma Very sweetly perfumed and rather spicy.
Flowers Pollination group 4.
Leaves Medium to large. Acute to narrow acute. Serrate. Medium thick. Sometimes slightly upward-folding. Very slightly undulating. Mid green. Undersides quite downy.

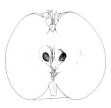

WARNER'S KING

Season   Late September to February
Picking time   Mid to end of September

This very old cooking apple, was known around the late 1700s as King Apple and it was under this name that Mr Warner, a small nurseryman of Gosforth near Leeds gave it to Mr Rivers of Sawbridgeworth, who re-named it Warner's King. It is said to have originated in an orchard in Weavering Street, Maidstone in Kent. Today trees are available from a few specialist nurseries. It is a very good, heavy cropping variety with a long season. The trees are vigorous, upright-spreading and produce spurs very freely. They are triploid, fairly hardy and suitable for growing in the North but prone to canker and scab. The cropping is heavy with large fruits which have a reasonably good, very acid flavour. They cook to a very fine fluff.

Size Very large, 95 × 76mm (3¾ × 3″).
Shape Flat-round to slightly conical. Broad and flattened at base. Often lop-sided. Well-rounded ribs. Irregular.
Skin Yellowish-green (144B) becoming yellow (8A) to slightly deeper yellow (12A) nearest the sun. Sometimes a slight flush of pinkish-brown (179B) or purplish-brown (176B). There can be some small patches and dashes of grey-brown russet. Lenticels numerous and fairly conspicuous greenish-white dots surrounded by a small circle of darker green, or pinkish-brown russet dots. Skin smooth and dry becoming greasy if stored.
Stalk Medium thick (3mm). Medium length (17–20mm). Frequently set at an angle.
Cavity Wide and deep. The ribs can enter the cavity. Lined with fine ochre-brown or grey russet which can scatter over base. Often lipped.
Eye Medium size. Closed or partly open. Sepals very long and slender, erect convergent with some tips reflexed. Fairly downy.
Basin Rather narrow. Medium to moderately deep. Slightly puckered.
Tube Cone or funnel-shaped.
Stamens Median sometimes towards basal.
Core line Median sometimes towards basal.
Core Median. Abaxile.
Cells Roundish ovate.
Seeds Acute. Fairly large, rather angular and curved Irregular. Often very thin.
Flesh White tinged slightly green. Rather coarse-textured. Crisp and juicy.
Aroma Nil.
Flowers Pollination group 2. Triploid.
Leaves Large. Acute. Either long narrow acute and sharply serrate, or bluntly serrate. Fairly thick and leathery. Slightly upward-folding and undulating. Dark green. Downward-hanging. Very downy.

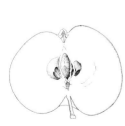

NORFOLK ROYAL

Season   Late September to early December
Picking time   Early September

This mid season dessert apple arose to notice in 1908 from a chance seedling at Wright's Nurseries at North Walsham in Norfolk. It was introduced in 1928 and named in 1930. At one time it was well known in the Eastern Counties but today it is only listed by one or two specialist nurseries. It is a very attractive exhibition variety. The trees are moderately vigorous, upright in habit and partial tip-bearers. They are fairly hardy and suitable for the colder areas. The cropping is very good and the fruits have a pleasant though not strong flavour.

Size Medium to large, 70 × 67mm (2¾ × 2⅝″).
Shape Conical to long-conical. Symmetrical or slightly lop-sided. Slight well-rounded ribs. Can be rather flat-sided towards apex. Fairly regular.
Skin Very pale yellow (8B) to pale whitish-yellow (2D) in parts. Quarter to almost completely flushed with brilliant red (45A) sometimes becoming crimson (46A) nearest the sun. Frequently flecked, dotted and striped with red (46B) over the yellow skin on shaded side. Indistinct broad, broken stripes of deep brownish-crimson (185A). Lenticels inconspicuous as small whitish-ochre or greenish-ochre dots. Skin smooth and very shiny. Becomes greasy if stored.
Stalk Fairly slender (2.5mm). Short to medium (8–15mm). Usually about level with base of fruit.
Cavity Narrow and deep. Lined with varying amounts of grey-brown russet over ochre-green which can streak out over base.
Eye Fairly small. Closed. Sepals rather small, sometimes slightly separated at base, erect and slightly convergent with tips reflexed. Very downy.
Basin Medium width and depth. Ribbed and slightly puckered. Some ochre-brown or light grey-brown russet may be present radiating from eye.
Tube Cone-shaped occasionally funnel-shaped.
Stamens Basal. Median towards basal when funnel-shaped tube.
Core line Prominent. Almost basal, clasping.
Core Median to sessile on long conical fruits. Axile.
Cells Round.
Seeds Obtuse. Large wide and rather flat. Blunt. Straight or slightly curved.
Flesh Creamy white tinged pink near the skin. Moderately firm but not hard. Fairly crisp and juicy. Rather coarse-textured.
Aroma Slight, pleasantly sweet.
Flowers Pollination group 5.
Leaves Medium size. Acute or narrow acute. Crenate or bluntly serrate. Medium thick. Upward-folding. Mid yellowish-green. Some rather downward-hanging. Undersides fairly downy.

GREENSLEEVES

Season   Late September to mid November
Picking time   Mid to late September

This is a new dessert apple from East Malling Research Station which received the Award of Merit from the RHS in 1981. It was raised in 1966 from James Grieve × Golden Delicious by Dr Alston. It is a mid season Golden Delicious type and could be used as a substitute for French Golden Delicious. It may be useful as a pollinator for some of the major varieties. Greensleeves makes a fairly upright compact tree of weakish moderate vigour, making it easy to manage. It is hardy and suitable for growing in the colder areas and will set a reasonable crop if self-pollinated. It is a partial tip-bearer. The cropping is very heavy and the trees bear fruit very early in life. The fruits are crunchy and sweet with a nice tangy bite in September and October but they go rather soft in November. The flavour is good but it soon fades and the skin is a little tough. Trees are widely available from nurserymen.

Size Medium, 67 × 60mm (2⅝ × 2⅜″) or 64 × 57mm (2½ × 2¼″).

Shape Round to oblong. Rounded at base, flattened at apex. Occasionally lop-sided. Slightly five-crowned at apex. Sometimes a hint of ribs. Regular.

Skin Pale green (145A or slightly paler), becoming rather whitish-yellow (5C). There can be a gentle flush of greyish-orange (170C–168C). Some very small patches of golden-brown russet. There can be a russetted hair-line. Lenticels conspicuous as grey-brown russet dots which are slightly raised giving the surface a slightly rough feel. Skin dry.

Stalk Fairly slender (2.5mm). Long (20–22mm). Extends well beyond base. Can be set at an angle.

Cavity Medium depth and width. Some grey-brown or ochre russet which can spread a little over base.

Eye Medium to quite large. Closed or slightly open. Sepals very long, narrow, erect convergent with tips reflexed. Stamens often visible. Downy.

Basin Moderately deep and fairly wide. Slightly beaded and slightly ribbed. There can be a smattering of golden-brown russet.

Tube Funnel-shaped, rather narrow.

Stamens Median.

Core line Median.

Core Median. Axile.

Cells Obovate.

Seeds Acute. Large. Fairly plump. Slightly curved.

Flesh Creamy white. Slightly coarse-textured. Juicy.

Aroma Very slight.

Flowers Pollination group 3.

Leaves Medium size. Broadly acute. Bluntly and broadly serrate. Thick and leathery. Upward-folding. Flat not undulating. Mid Yellowish-green. Undersides not very downy.

PEASGOOD NONSUCH

Season   Late September to December
Picking time   Mid September

This handsome, highly esteemed, culinary apple was raised in England by Mrs Peasgood of Stamford in Lincolnshire from a seed said to have been from the Catshead Codlin, sown in 1858. It was awarded a First Class Certificate from the RHS in 1872. It has always been primarily a garden and exhibition variety, being too soft for commercial use and trees are fairly widely available from nurserymen today. The trees are moderately vigorous, spreading in habit and produce spurs fairly freely. The flavour of the fruit is extremely good, being slightly acid and yet quite sweet, and they make very good baking apples. The cropping is heavy with large fruits.

Size Large, 83 × 70mm (3¼ × 2¾″) to very large, 92 × 76mm (3⅝ × 3″).

Shape Round slightly flattened to slightly conical. Flattened at base and apex. Symmetrical or sometimes a little lop-sided. Regular. Not ribbed.

Skin Pale yellowish-green (150C) becoming pale yellow (8C–6C) deepening to gold (13B) or orange-yellow (22A) nearest the sun. Quarter to three quarters overlaid with short broken stripes of bright red (45A). There are usually small, slightly scaly, ochre-brown russet patches. Lenticels fairly conspicuous as very small ochre-brown, green or pinky-red dots. Skin smooth and very slightly greasy and soft to the touch.

Stalk Fairly stout (3.5mm) to stout (4mm). Short (8mm).

Cavity Wide and fairly deep. Green. Some grey-brown russet present which can streak out over base.

Eye Medium size. Partly or completely open. Sepals erect or convergent with tips reflexed. Fairly downy.

Basin Fairly deep and wide. Even. Sometimes very slightly puckered.

Tube Long funnel-shaped, broad at eye.

Stamens Median.

Core line Median, at junction of bowl and tube of funnel.

Core Median. Axile.

Cells Obovate.

Seeds Acute. Oval pointed, rather tufted. Fairly plump.

Flesh Yellowish, tinged slightly green. Fairly juicy. Tender.

Aroma Quite strong, sweetly scented.

Flowers Pollination group 3.

Leaves Fairly large. Broadly oval almost roundish. Bluntly pointed serrate. Thick and leathery. Mostly flat. Mid, fairly bright rather yellowish green. Undersides slightly downy.

— EMPEROR ALEXANDER —

Season   September to November
Picking time   Mid September

This old cooking apple, sometimes listed as Alexander, originated from Russia. It was known in the 1700s but not introduced to England until the early 1800s. It was imported from Riga in Russia by Messrs Lee and Kennedy in 1817, who exhibited examples at the London Horticultural Society in that year. According to B. Maund in *The Fruitist* 1845–51, the apple is a native of the southern provinces of Russia. It was named after the Emperor of Russia to whom apples were said to be sent annually as a present. It is a handsome exhibition variety but is now seldom listed. The trees are vigorous, upright-spreading, partial tip-bearers. The cropping is moderate. The fruits are sub-acid with no particular flavour and they cook to a pale yellow fluff.

Size Very large, 92 × 76mm (3⅝ × 3″).
Shape Round-conical to conical. Very broad at base. Can be slightly five-crowned at apex. Usually symmetrical. Rather angular and flat-sided, with slight sometimes angular ribs. Fairly regular.
Skin Pale yellowish-green (between 145B and 150C). Quarter to three-quarters flush with mottled orangey-red (42B). Broad broken stripes and splashes of red (45A). Lenticels inconspicuous ochre russet dots on flush becoming smaller and more numerous pinky-white dots towards apex, and small grey-brown russet dots on green skin. Skin smooth and dry becoming greasy.
Stalk Fairly stout (3.5mm) and medium length (18mm). Level with base or slightly beyond.
Cavity Wide and deep. Regular. Lined with scaly light brown or grey-brown russet encircling cavity and sometimes spreading slightly over base.
Eye Medium to large. Half open. Sepals erect with tips reflexed. Downy.
Basin Deep and even. Narrow to medium width. Slightly ribbed.
Tube Cone-shaped or funnel-shaped.
Stamens Median.
Core line Median or basal.
Core Median to sessile. Axile, sometimes open.
Cells Ovate. Can be slightly tufted.
Seeds Obtuse. Short and roundish. Medium size. Fairly plump.
Flesh White with slight greenish-yellow tinge. Coarse-textured. Dry. Firm.
Aroma Nil, slightly acidic when cut.
Flowers Pollination group 3.
Leaves Medium size. Broadly acute. Broadly crenate. Medium thick. Slightly upward-folding and slightly undulating. Mid green. Downy.

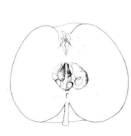

— STIRLING CASTLE —

Season   September to December
Picking time   Mid September

This is an old Scottish cooking apple which was raised at Stirling in about 1830 by John Christie, a nurseryman at Causewayhead on the road to the Bridge of Allan. It was introduced by Messrs Drummond of Stirling and first recorded in 1831. The trees are spreading, with weak growth, and are spur-bearers. They are hardy and suitable for growing in the colder areas. The fruit is acid, juicy and of good flavour when tried in September but apparently it improves if left until October. I found it juicy and full flavoured when eaten as a dessert.

Size Medium, 67 × 57mm (2⅝ × 2¼″) to medium large, 76 × 60mm (3 × 2⅜″).
Shape Flat-round. Flattened at base and apex. Symmetrical. Sometimes a hint of well-rounded ribs. Regular.
Skin Bright yellowish-green (144C) becoming pale yellow (154C). Some fruits lightly flushed with rather streaky, milky pink (35C) or pinky-orange (34D). Varying amounts of streaked or speckled scarf skin chiefly at base. Lenticels fairly conspicuous as green or brown russet dots sometimes surrounded by a circle of pink on flush. Skin smooth and shiny becoming greasy.
Stalk Variable. Medium (3mm) to stout (4mm). Medium length (12–20mm). Extends beyond base.
Cavity Deep and wide. Usually varying amounts of ochre-brown russet which can streak a little over base.
Eye Medium size. Partly open. Stamens fairly broad based, some slightly separated at base, erect convergent with tips reflexed but frequently broken off. Slightly downy.
Basin Deep and wide. Usually russet free but occasionally a small amount of grey-brown russet.
Tube Cone-shaped.
Stamens Median.
Core line Median.
Core Median. Abaxile, or axile.
Cells Roundish, slightly obovate or slightly ovate.
Seeds Obtuse. Roundish or oval. Blunt. Fairly plump and straight.
Flesh White. Rather coarse-textured and soft. Fairly juicy.
Aroma Nil.
Flowers Pollination group 3.
Leaves Medium size. Broadly acute. Serrate. Medium thick. Slightly upward-folding, slightly undulating. Mid green. Undersides downy.

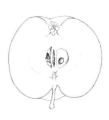

Season   October to December
Picking time   Late September

The origin of this apple appears to have been lost. It is thought to have probably originated in England and was first recorded here in 1872. It has been catalogued by most nurseries since the early part of this century and is now the most important commercial russet in the United Kingdom, as demand for a russet apple has encouraged recent planting. The young tree crops regularly and produces a good sized fruit but the mature tree tends to be a little biennial in cropping. The trees have moderate vigour, are upright in habit and are spur-bearers, producing spurs extremely freely. The fruit is sweet with a rich nutty flavour, but the skin is rather tough. Trees are widely available from nurserymen and are reasonably hardy and suitable for growing in the northern and western parts of the country. It received the Award of Merit from the RHS in 1980.

Size Medium 64 × 48mm (2½ × 1⅞″). 67 × 57mm (2⅝ × 2¼″).
Shape Flat-round. Well flattened at base and apex. Symmetrical or a little lop-sided. Usually no ribs but occasionally a slight hint of one. Regular.
Skin Yellowish-green (150B) becoming golden-yellow (7A). Often up to half flushed with brownish-ochre (163A) to light brownish-red (171B). Half or more covered with fine ochre (approx: 163B) to grey-brown (199B) russet. Lenticels very conspicuous whitish dots becoming larger towards base. Skin very dry.
Stalk Medium thick (3mm) and very short (6–10mm). Sunk well down within cavity.
Cavity Medium depth to rather shallow. Narrow. Frequently greenish-yellow with the russet rather thinly spread and skin beneath showing through, particularly at shoulder.
Eye Quite large, usually wide open. Sepals erect or slightly convergent with tips reflexed, exposing stamens.
Basin Medium width and depth. Sometimes very slightly ribbed or puckered. Regular.
Tube Funnel-shaped.
Stamens Median towards basal.
Core line Almost basal.
Core Median. Axile.
Cells Small, roundish ovate to oval.
Seeds Obtuse. Oval, very blunt.
Flesh Cream tinged slightly yellow. Crisp and firm. Fine-textured. Not very juicy.
Aroma Nil.
Flowers Pollination group 2.
Leaves Medium size. Acute. Very sharply and broadly serrate. Flat not undulating, sometimes slightly upward-folding. Thin. Mid green. Slightly downy.

Season   September to January
Picking time   Late September

This is a new mid season to late cooking apple produced by the East Malling Research Station in Kent, England. It was derived from a seedling of Cox's Orange Pippin pollinated by Lane's Prince Albert. Bountiful has several advantages over the Bramley in that it is a diploid, thus requiring only one pollinator. The trees are of compact habit, making them very suitable for small gardens. Bountiful makes cordons easily and is suitable for growing in tubs. It has considerable resistance to apple mildew. The cropping is very heavy and the fruits have a good flavour though not particularly outstanding. They are sub-acid and require no additional sugar and cook to a yellow fluff. Bountiful can be eaten as a dessert apple in late winter. Trees are available from some fruit tree nurserymen.

Size Large, 76 × 57mm (3 × 2¼″).
Shape Round to round conical, flattened at base. Slight well rounded ribs and can be rather flat-sided. Fairly symmetrical but can be slightly lop-sided. Fairly regular.
Skin Yellowish green (151D) becoming yellow (12A). Can be up to a quarter flushed with brownish-orange (171B) with broken scarlet stripes and flecks (42A). Lenticels inconspicuous purplish-grey dots surrounded by light yellowish-white on yellow skin and light brownish-orange on flush. Fine grey-brown russet flecking around apex. Skin smooth and dry.
Stalk Medium thick (3mm). Medium length (20mm). Protrudes slightly beyond base. Can be set at an angle.
Cavity Broad and deep. Lined with ochre-green overlaid with light grey-brown russet which can streak a little over base. Quite pronounced large dots of yellow skin can show through the russet.
Eye Medium size. Partly open with sepals erect and tips reflexed or broken off. Fairly downy.
Basin Medium width and fairly deep. Slightly ribbed.
Tube Cone-shaped.
Stamens Marginal or median.
Core line Basal, clasping.
Core Median or slightly sessile. Axile slightly open.
Cells Obovate.
Seeds Acute. Regular. Fairly plump.
Flesh Creamy-white tinged yellow. Firm. Fine-textured. Fairly juicy. Fairly tender.
Aroma Sweetly scented but not strong.
Flowers Pollination group 3.
Leaves Medium size. Acute. Crenate. Medium thick. Upward-folding. Flat not undulating. Mid grey-green. Undersides very downy.

Season Mid October to December
Picking time Early October

The origin of this old dessert apple seems somewhat confused. It is thought to be of English origin and introduced in the early 1800s by Mr Kirke, a nurseryman of Brompton, who named it King of the Pippins. An older name for this apple was Golden Winter Pearmain. It was planted extensively in the past for commercial use, though not much during this century. Trees are available from a few specialist nurserymen. The trees are moderately vigorous, upright in habit and produce spurs very freely. They are reasonably hardy and are also suitable for growing in the west country. The fruits are sweet crisp and juicy with a very rich and vinous, rather nutty flavour. A distinctive and lovely apple.

Size Medium, 67 × 64mm (2⅝ × 2½") to 60 × 57mm (2⅜ × 2¼").

Shape Oblong-conical. Frequently a little lop-sided. Slight trace of well-rounded ribs more evident at apex where it may be five-crowned. Regular.

Skin Slightly greenish-yellow (150B) becoming yellow (12B) sometimes with some green stripes or patches. Quarter to three quarters flushed with brownish-orange (169B) with short broken stripes, and the occasional long stripe of bright red (45A). Some small pale grey russet flecks and patches. Lenticels conspicuous as green or whitish-ochre russet dots on yellow skin and whitish-russet dots on flush. Skin smooth and almost dry.

Stalk Variable. Either fairly slender (2.5mm) and quite long (22mm) or stout (4mm) and rather short (10mm). Usually extends beyond base.

Cavity Medium width and medium depth. Lined with fine greenish-ochre russet which can streak and scatter over base. Cavity bright green where not covered with russet.

Eye Large. Partly open. Sepals broad based and very long. Connivent with tips well reflexed. Fairly downy.

Basin Rather shallow and wide. Slightly ribbed.

Tube Funnel-shaped.

Stamens Median.

Core line Basal, clasping.

Core Median. Axile open.

Cells Obovate.

Seeds Obtuse or acute. Round and plump.

Flesh Creamy-white. Firm and crisp. Fine-textured. Juicy.

Aroma Slight, sweetly scented.

Flowers Pollination group 5. Biennial.

Leaves Medium to small. Oval. Serrate. Medium thick. Flat. Mid blue-green. Downy.

Season October to November
Picking time Late September

This is an old American mid season dessert apple. It originated at Bolton, Worcester County, Massachusetts and was first recorded in 1844. The *Herefordshire Pomona* records that it was Mr Rivers of Sawbridgeworth who introduced it to England in the early 1800s. It is noted for its good flavour, sweet and aromatic, but needs full sun to develop its full potential. The trees are moderately vigorous, very upright in habit and produce spurs freely. It is suitable for growing in the West Country and trees are widely available from nurserymen. The cropping is irregular.

Size Medium, 64 × 60mm (2½ × 2⅜").

Shape Long-conical. Rounded at base. Rather indistinct broad ribs. Can be flat-sided. Sometimes symmetrical but often lop-sided. Slightly irregular.

Skin Greenish-yellow (1B) becoming yellow (12B). Quarter to three quarters flushed with dull orange-red (42B) to a deeper dull red (45C) nearest the sun. Rather indistinct fairly long narrow stripes of crimson-red (46A). Some small greyish-ochre russet patches. Lenticels indistinct numerous tiny grey-brown or reddish-russet dots. Skin smooth and dry.

Stalk Slender (2mm). Medium length (13–17mm). Usually protrudes slightly beyond base.

Cavity Fairly wide to rather narrow. Medium to fairly deep. Frequently lipped on one side.

Eye Small. Closed or partly open. Sepals erect convergent with tips reflexed. Slightly downy.

Basin Shallow and rather small. Slightly ribbed and puckered, sometimes a little beaded. There can be a small amount of brown russet present.

Tube Tiny, funnel-shaped.

Stamens Almost marginal.

Core line Median towards basal.

Core Median. Abaxile.

Cells Elliptical. Tufted.

Seeds Acuminate or acute. Numerous. Plump.

Flesh Creamy-white tinged green towards core. Firm but tender. Coarse-textured. Skin rather tough.

Aroma Slight, sweetly scented.

Flowers Pollination group 5.

Leaves Medium size. Acute. Serrate. Medium thick. Flat not undulating. Slightly upward-folding. Mid grey-green. Undersides slightly downy.

GRAVENSTEIN

Season   Mid September to December
Picking time   Late August

There are many theories regarding the origin of this old dual-purpose apple. Some say it was found at Castle Grafenstein, Schleswig-Holstein, in Germany, others that it was sent there from Italy or Southern Tyrol as 'Ville Blanc,' or possibly that scions were sent from Italy by a brother of Count Chr. Ahlefeldt's of Graasten Castle, South Jutland. It was thought to have arrived in Denmark about 1669 and in London in 1819. The trees are vigorous, upright-spreading and are unsuitable for the small garden. They are spur and tip bearers and triploid. The fruits are sweet with a good acid balance and distinctive flavour. They make a very nice large dessert and also cook to a fine juicy fluff. The cropping is fairly good. Trees are available from some nurseries.

Size Medium large, 73 × 67mm (2⅞ × 2⅝″) to large 83 × 70mm (3¼ × 2¾″).

Shape Oblong. Large well pronounced ribs running from base to apex. Rather five-crowned at apex. Can be flat-sided. Often lop-sided. Irregular.

Skin Yellow-green (145B) to greenish-yellow (150B) becoming pale yellow (5C). Quarter to three quarters flushed with rather thin orangey-red (171C) which appears in broad bands from base to apex rather than one solid patch. Sparsely striped with greyish-red to crimson-red (46A). Some scarf skin at base. Lenticels conspicuous greyed-purple or pinky dots on flush and green dots on yellow skin. Skin very smooth and dry, becoming greasy.

Stalk Stout (4mm) and short (10mm). Sometimes with fleshy knob. Within cavity.

Cavity Deep. Medium width. Partly or more lined with fine ochre which can ray out.

Eye Large. Either closed with sepals connivent or open with sepals erect and tips reflexed. Very downy with the down extending on to basin.

Basin Medium to wide. Ribbed. Sometimes beaded.

Tube Funnel-shaped or wide cone-shaped.

Stamens Median when funnel-shaped, towards basal when cone-shaped.

Core line Indistinct. Median to almost basal.

Core Somewhat distant. Abaxile.

Cells Elliptical or round.

Seeds Acute. Rather small, sparse. Plump. Curved.

Flesh White tinged yellow near the skin. Rather coarse-textured. Firm. Fairly crisp. Very juicy.

Aroma Fairly strong, sweet and fruity.

Flowers Pollination group 1. Triploid.

Leaves Fairly large. Broadly acute. Serrate. Medium to thin. Flat not undulating. Dark green. Downy.

JONAGOLD

Season   November to February
Picking time   Mid October

This fairly new introduction is becoming important commercially in the UK. It is a large, late season dessert apple and originates from the New York State Agricultural Experiment Station in Geneva, New York, where it was raised in 1943 from Golden Delicious × Jonathan. It first fruited in 1953, and was introduced in 1968. It received the Award of Merit from the RHS in 1987. The fruits are often very poorly coloured and there are a number of coloured sports coming on to the market – for example, Red Jonagold. Trees are triploid, very vigorous, wide-spreading and produce spurs very freely. They are not suitable for growing in a small garden or restricted form unless grafted on to a dwarfing rootstock. The fruits are sweet with a good rich flavour. The skin is a little tough. Trees are available from several fruit tree nurseries.

Size Large, 77 × 64mm (3 × 2½″), or larger on young trees.

Shape Round. Base slightly flattened, shoulders often rounded. Flattened at apex. Slightly ribbed, more noticeable and angular towards apex. Slightly five-crowned at apex. Symmetrical or lop-sided. Regular.

Skin Light yellow-green (145A) becoming greenish-yellow (151A). Slightly to half flushed and mottled bright red (42A) with short broken stripes. Occasional small patches of grey-brown russet. Lenticels fairly conspicuous grey-brown russet dots which look slightly raised. Surface dry and bumpy.

Stalk Fairly long to long (20–30mm). Fairly slender to medium (2.5–3mm). Protrudes beyond base.

Cavity Very deep and wide. Partly or more lined with fine ochre or grey russet with a silvery sheen, which scatters over base. Lenticels enter cavity.

Eye Smallish for size of fruit. Slightly open. Sepals erect and convergent but not touching, with some tips slightly reflexed. Fairly downy.

Basin Fairly wide and deep. Slightly ribbed and puckered. Some slight cinnamon-brown russet.

Tube Cone-shaped.

Stamens Basal.

Core line Basal clasping, often following cell outline.

Core Median. Axile.

Cells Obovate.

Seeds Acuminate. Fairly plump or rather starved. Rather angular. Straight or slightly curved.

Flesh Creamy white. Firm. Juicy. Fine-textured.

Aroma Nil, slightly woody when cut.

Flowers Pollination group 4. Triploid.

Leaves Medium size. Acute to broadly acute. Sharply and finely serrate. Medium thick. Flat. Upward-folding. Light greyish-green. Downy.

CHARLES ROSS

Season   October to December
Picking time   Mid September

This handsome dual purpose apple was raised from Peesgood Nonsuch × Cox's Orange Pippin by Charles Ross, gardener to Capt. Carstairs at Welford Park in Berkshire from 1860–1908. This apple was originally named Thomas Andrew Knight who was president of the RHS. First exhibited in 1890, it received an Award of Merit in 1899. In that year at Capt. Carstair's request, the name was changed and the apple received a First Class Certificate as Charles Ross. It is grown on a small scale commercially in the UK and trees are widely available from nurseries. The trees are moderately vigorous, upright-spreading and produce spurs very freely. They are fairly resistant to scab but prone to Capsid bug and suitable for the North and West. The fruits are juicy, sweet and flavoursome but become flat and flavourless by late October. The flesh is yellow when cooked and stays fairly intact, sweet with reasonable flavour.

Size Large, 80 × 70mm (3⅛ × 2¾″).
Shape Round, slightly conical. Flattened at base and apex. Usually symmetrical but can be a little lop-sided. No ribbing. Regular. A very heavy apple.
Skin Greenish-yellow (151D). Half to three quarters flushed orange-red (34A) to ochre-orange away from the sun (168B). Broad broken stripes of red (45A) often overlaid with greyed scarf skin. Scarf skin at base and on cheeks. Lenticels conspicuous ochre-russet dots on flush, grey-brown russet dots on yellow skin. They can be angular or star-shaped. Some russet dots and patches especially towards base. Numerous whitish dots at apex. Skin fairly smooth, becoming greasy.
Stalk Stout to very stout (4–6mm). Short (10mm). Often fleshy. Usually sunk well within cavity.
Cavity Wide. Medium depth. Usually green. Lined with grey, or grey-brown slightly scaly russet. Scarf skin can be cavity. Sometimes lipped.
Eye Medium size. Open or partly open. Sepals erect convergent with tips reflexed. Very downy.
Basin Medium depth to shallowish. Fairly wide. Even. Can be slightly ribbed.
Tube Funnel-shaped, rather long and narrow.
Stamens Median or towards marginal.
Core line Median joining at funnel neck.
Core Median. Axile.
Cells Roundish obovate. Rather small.
Seeds Obtuse. Fairly plump, regular and straight.
Flesh Creamy white. Crisp and juicy. Rather coarse.
Aroma Very slight. Sweetly perfumed.
Flowers Pollination group 3.
Leaves Medium size. Acute. Crenate. Medium thick. Slightly upward-folding and slightly undulating. Mid yellowish-green. Slightly downy.

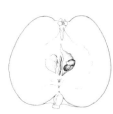

COX'S POMONA

Season   October to December
Picking time   Early September

This apple was raised in England in about 1825 by Richard Cox of Colnbrook Lawn, Slough, Buckinghamshire. It is a mid to late season culinary or possibly dual purpose apple, said to be a seedling of Ribston Pippin possibly crossed with Blenheim Orange. It was introduced by Mr Smale of Colnbrook Nursery in Slough. It is not grown commercially but listed by one or two specialist fruit tree nurserymen. The trees are moderately vigorous, upright-spreading, spur-bearers. The cropping is good. As a dessert apple I thought it crisp, juicy and good. When cooked it has a fine, delicate flavour, sub-acid and not requiring any additional sugar and it stays fairly intact.

Size Large, 83 × 64mm (3¼ × 2½″).
Shape Flat-round to round-conical. Well flattened at base and apex. Well pronounced ribs. Five-crowned. Symmetrical or lop-sided. Irregular.
Skin Pale yellow-green (145A) to pale greenish-yellow (1C). Quarter to three quarters flushed with brilliant red (46A–46B) fading at the edges to a paler red (47A) where there are prominent broad stripes and speckles of the brighter red (46A–46B). In the denser part of the flush nearest the sun the stripes are less prominent and crimson (53A) in colour. Lenticels not very distinct small pinky-white dots on flush but they are more numerous at the base where they are more elongated in shape. A brilliant and attractive fruit. Skin smooth and very slightly greasy, becoming more greasy if stored. Very shiny.
Stalk Variable. Stout (4mm) and short (10mm) or fairly slender (2.5mm) and longer (15–20mm). Very deeply sunk into cavity.
Cavity Deep and wide. Usually green or yellowish and sometimes lined with grey-brown russet.
Eye Medium. Open. Sepals short, separated at base, erect with tips often broken off. Fairly downy.
Basin Deep and rather narrow. Irregular. Ribbed.
Tube Deep cone-shaped.
Stamens Median. Axile.
Core line Median, appearing to pull the sides of the tube apart.
Core Median.
Cells Roundish-obovate.
Seeds Obtuse. Rather broad. Fairly plump.
Flesh White tinged very slightly green. Rather coarse-textured. Firm but tender.
Aroma Slight, sweetly fragrant.
Flowers Pollination group 4.
Leaves Medium size. Acute. Crenate. Medium thick. Slightly upward-folding and very slightly undulating. Fairly dark greyish-green. Very downy.

McINTOSH RED

Season   October to December
Picking time   Mid September

This mid to late dessert apple is of Canadian origin. It was discovered in 1796 by John McIntosh at Matilda Township, Dundas County, Ontario, said to be a chance seedling. It was propagated by Allan McIntosh and named in about 1870. It is widely grown in Canada and the United States today. There is a race of McIntosh-type apples, including Tydeman's Early Worcester and Vista Bell, which are noted for their highly perfumed smell and taste, and there are many highly coloured clones in existence. The trees are moderately vigorous, spreading and produce spurs very freely. They are however susceptible to canker. The cropping is good and the fruits sweet with a strong vinous flavour. Listed by one or two specialist nurseries.

Size Medium large, 70 × 67mm (2¾ × 2⅝″).
Shape Round to flat-round with fairly well defined ribs which are slightly more noticeable towards apex. Can be five-crowned at apex. Symmetrical or sometimes a little lop-sided. Slightly irregular.
Skin Yellow-green (144C) becoming whitish yellow (3C to 3D). Half to almost entirely flushed with crimson (53A) to brownish-crimson (185A) on well coloured fruit. Short broken purplish-crimson stripes (187B) which are lighter but more noticeable over paler flush. The fruit is covered with fine lilac bloom which rubs off when touched and the fruit polishes to a high shine. Lenticels fairly noticeable as small yellow or whitish dots. Skin smooth and dry.
Stalk Slender to medium (2–3mm). Fairly short (10–20mm). Level with base or slightly beyond.
Cavity Medium width and fairly deep. Can be slightly ribbed. Partly lined with greenish-ochre or slightly scaly grey-brown russet.
Eye Very small. Tightly closed or slightly open. Sepals small, erect convergent. Very downy.
Basin Rather small. Medium depth and rather narrow. Slightly ribbed with some beading.
Tube Conical.
Stamens Median.
Core line Basal, clasping or towards median.
Core Median. Abaxile.
Cells Round.
Seeds Fairly large. Acute. Plump sometimes angular.
Flesh Very white tinged pink near the skin. Fine-textured, tender, almost soft. Very juicy.
Aroma Strong and very sweetly perfumed.
Flowers Pollination group 2.
Leaves Medium size. Acute. Bluntly serrate or crenate. Medium thick. Flat. Slightly upward-folding. Light yellowish-green. Very downy.

LORD DERBY

Season   October to December
Picking time   Late September

This well known mid to late culinary apple was raised in England by Mr Witham, a nurseryman of Stockport in Cheshire. It was first recorded in 1862. It is grown on a medium scale commercially in the UK, being a good variety for marketing before Bramley's Seedling. It is a prolific and regular cropper but may require a little thinning to obtain large fruits. The fruits should be marketed while they are green, before they turn yellow. The trees are moderately vigorous, upright-spreading, show a good resistance to scab and succeed well on wet soils. They are very hardy and suitable for the North. The trees produce spurs freely and the flowers show a good degree of self-fertility. The fruits are light in weight, sub-acid with good flavour and stay intact when cooked. Trees are available from nurserymen.

Size Large, 83 × 70 mm (3¼ × 2¾″).
Shape Round-conical to oblong-conical. Pronounced rather angular ribs. Often flat-sided, or slab-sided. Flattened at base, five-crowned at apex. Can be a little lop-sided. Irregular.
Skin Strong, bright, grass green (144A) becoming bright yellow (12A). Much overlaid with scarf skin that starts at the base and extends partly up the cheeks making the colour of the skin a rather milky green (between 144D and 143D). Occasionally a slight blush of pinkish-brown (176C) nearest the sun. Lenticels fairly conspicuous whitish-green dots, smaller and more numerous towards apex. Skin smooth and dry.
Stalk Very stout (6mm), often fleshy and very short (5–6mm). Set well within cavity.
Cavity Wide and fairly deep. Green with some streaky scarf skin. Occasional tiny patch of very fine pinky-brown russet next to stalk.
Eye Medium size. Closed or partly open. Sepals broad-based connivent with tips often reflexed right back. Downy.
Basin Deep. Angular, often very pinched looking. Much ribbed and puckered. Sometimes beaded.
Tube Deep cone-shaped or funnel-shaped.
Stamens Median or towards marginal.
Core line Median.
Core Median. Abaxile.
Cells Roundish ovate, or rather elliptical.
Seeds Obtuse to acute. Regular not curved.
Flesh Greenish-white. Slightly coarse-textured. Rather dry. Soft.
Aroma Pleasant, slightly acid not strong.
Flowers Pollination group 4.
Leaves Medium size. Acute. Serrate. Medium to rather thin. Upward-folding very slightly undulating. Mid blue-green. Undersides very downy.

Season October to December
Picking time Late September

A mid season culinary apple which was raised by John Graham of Hounslow, England. The parentage is unrecorded. It was first recorded in 1888 and was introduced by G. Bunyard & Co. of Maidstone, Kent in 1893. It is grown on a small scale commercially but trees have a limited availability from nurserymen. The trees are weak, upright-spreading and spur-bearers. The flesh is sub-acid, with a pleasant but weak flavour and remains intact when cooked. The cropping is good and regular, though the fruits tend to drop all at once, so it needs watching at harvest time. It is late flowering which makes it useful for growing in areas subject to late frosts.

Size Medium, 67 × 67mm (2⅝ × 2⅝″) or 70 × 54mm (2¾ × 2⅛″) or large, 83 × 73mm (3¼ × 2⅞″).

Shape Oblong to conical. Very irregular with large ribs some more prominent than others. Sometimes flat-sided especially towards apex. Usually rather lop-sided.

Skin Greenish-yellow (150B) becoming yellow (10A) with deeper yellow (14B) nearest the sun. There can be a slight flush of pale ochre-brown (164B) deepening to pinkish-brown (173A) nearest the sun. There can be some patchy scarf skin at or towards base. Lenticels conspicuous green or grey-brown russet dots, or whitish dots surrounded by a circle of greyed brown on the brighter flush. Some fruits have a light dusting of grey-brown russet. Skin smooth and slightly greasy, becoming more greasy.

Stalk Stout (4 mm). Short (10mm). Within cavity or level with base. Sometimes fleshy.

Cavity Fairly wide and medium depth. Partly lined with scaly golden-brown russet.

Eye Medium size. Closed or slightly open. Sepals broad based, erect and connivent. Tips reflexed or broken off. Moderately downy.

Basin Medium depth and fairly narrow. Extremely pinched. Can be some large fleshy beads.

Tube Funnel-shaped. Often extending into core.

Stamens Median.

Core line Median.

Core Median to distant. Abaxile.

Cells Obovate or elliptical.

Seeds Acuminate. Plump.

Flesh Creamy-yellow. Firm. Coarse-textured. Juicy.

Aroma Slight, sweetly aromatic.

Flowers Pollination group 5.

Leaves Medium size. Acute. Bluntly pointed. Rather thin. Upward-folding, occasionally downward-folding. Mid yellowish-green. Downward-hanging. Undersides fairly downy.

Season October to December
Picking time Late September to early October

A well known mid to late dual purpose apple which was raised in England by Mr Charles Ross of Newbury in Berkshire from Peasgood Nonsuch × Cox's Orange Pippin. It was introduced by Clibrans of Altrincham. It was first recorded in 1900 when it received the Award of Merit from the RHS. At one time it was grown commercially, but today trees are available in only one or two specialist nurseries. The trees are moderately vigorous, spreading in habit and spur-bearers. They are fairly hardy and are also suitable for growing in the western areas of the country. The cropping is good though it tends to be biennial. The fruits are slightly acid with reasonably good flavour. It is a very refreshing crunchy apple when eaten as a dessert and has a nice tang.

Size Medium large, 73 × 58mm (2⅞ × 2¼″).

Shape Flat-round. Well flattened at base and apex with very slight trace of well rounded ribs. Symmetrical or sometimes slightly lop-sided. Fairly regular.

Skin Pale yellowish-green (150B). Quarter to half flushed with bright but rather thin orangey-red (34B) with fairly distinct short broken stripes of bright red (42A). A delicate network of fine greyish-ochre russet present in varying amounts. Lenticels inconspicuous as greenish-white or grey-brown dots. Skin smooth and dry.

Stalk Medium to stout (3–4.5mm) and short (10mm). Sunk well within cavity.

Cavity Fairly wide and deep. Lined with fine ochre or green-brown russet which can scatter over base.

Eye Medium size. Partly or wide open. Sepals short, erect, slightly separate at base when eye is wide open, with tips reflexed. Stamens usually present. Fairly downy.

Basin Wide and deep. Very slightly ribbed. Small amounts of fine brown russet can be present.

Tube Funnel-shaped.

Stamens Median.

Core line Median.

Core Median. Axile.

Cells Round.

Seeds Acute. Rather roundish and pointed. Fairly plump. Straight.

Flesh Creamy-white. Very crisp and juicy. Fine-textured.

Aroma Almost nil.

Flowers Pollination group 3. Biennial.

Leaves Medium size. Oval. Bluntly serrate. Medium thick. Slightly upward-folding not undulating. Mid green. Undersides downy.

Season   October to December
Picking time   Early September

A mid season culinary apple of English origin that was raised by Mr Allan at Gunton Park in Norwich. It is thought to be from Warner's King × Waltham Abbey. It was first recorded in 1901 when it received the Award of Merit from the RHS and it received a First Class Certificate in 1902, when it was introduced.

The trees are reasonably hardy and well known in the eastern counties, as the name suggests. They make vigorous growth, are spreading in habit and are partial tip-bearers. The cropping is good and the flavour of the fruits is reasonable. When cooked it is pale cream in colour, acidic, and breaks up completely though not to a fluff. Trees have a very limited availability from specialist nurserymen today.

Size Large, 80 × 67mm (3⅛ × 2⅝″).

Shape Round to round-conical. Flattened at base. Symmetrical or lop-sided. Usually fairly distinct broad ribs which can be angular. Sometimes one rib is more pronounced than the others making it irregular, otherwise fairly regular. Can be flat-sided.

Skin Pale yellowish-green (145A) becoming pale whitish-yellow (2C). There can be a slight flush of pinkish-brown (176C). Lenticels usually fairly conspicuous white dots surrounded by an areola of green becoming smaller and more numerous towards the apex. Lenticels can also be rather indistinct pinkish-brown dots. Skin smooth and dry.

Stalk Medium (3mm) and short (8–10mm) usually level with base.

Cavity Narrow. Medium to shallow in depth. Lined with scaly brown russet which can streak out over base.

Eye Large. Open. Sepals erect convergent or almost flat convergent, separated at base. Fairly downy. Stamens often visible.

Basin Medium depth. Medium width, rather irregular being slightly ribbed and slightly puckered.

Tube Rather wide and shallow, cone-shaped.

Stamens Median.

Core line Basal clasping. Median if tube is funnel-shaped.

Core Median. Abaxile.

Cells Elliptical. Tufted.

Seeds Usually acuminate, some acute.

Flesh Creamy-white. Coarse-textured. Fairly juicy.

Aroma Nil, very slight when cut.

Flowers Pollination group 2.

Leaves Medium to large. Broadly acute. Finely serrate. Rather thin. Slightly upward-folding and undulating. Mid blue-green. Fairly downy.

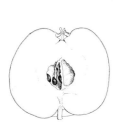

Season   October to December
Picking time   Late September

This mid to late dessert apple was raised in Lincolnshire, England by Thomas Laxton some time before 1884. It was exhibited initially as Brown's South Lincoln Beauty in 1889 by W. & J. Brown of Stamford and in 1894 received a First Class Certificate from the RHS under that name. In 1894 the name was changed to Allington Pippin and it received the Award of Merit. It was introduced by G. Bunyard & Co. in 1896. Earlier this century it was widely grown in Kent, Cambridgeshire and the Isle of Ely and is still widely listed. The trees are fairly hardy and are suitable for growing in the North and West. They are moderately vigorous and make a lot of young growth, are upright-spreading and spur-bearers. The cropping is heavy though it tends to be biennial. The fruits are rather dull looking though they have a rich aromatic flavour and are crunchy but not hard. Often confused with Laxton's superb.

Size Medium large, 70 × 61mm (2¾ × 2⅜″).

Shape Conical. Slight trace of well-rounded ribs. Slightly lop-sided or symmetrical. Regular.

Skin Pale whitish-green (145B) to lemon yellow (5C) becoming pale yellow (8B). Quarter to three quarters flushed with greyed-red (179A) or greyed orange (173A). A few indistinct broken stripes of deeper greyish-red (181A). Much overlaid with scarf skin giving it a bloomed appearance. Lenticels are conspicuous grey-brown russet dots, sometimes star-shaped. Some grey-brown or greyish-ochre russet patches. Skin smooth and dry.

Stalk Medium to fairly stout (3–3.5mm). Fairly short (15mm). Protrudes slightly beyond base or level.

Cavity Medium width. Medium to fairly deep. Regular. Some grey-green, slightly scaly russet streaks and scatters from cavity out over shoulder.

Eye Medium size, partly or fully open. Sepals very long and tapering, erect or convergent with tips well reflexed. Not very downy.

Basin Medium depth Fairly wide. Slightly ribbed. With occasional patch of grey-brown russet.

Tube Funnel shaped.

Stamens Basal.

Core line Basal, clasping.

Core Median. Axile, closed or open.

Cells Ovate.

Seeds Numerous. Acute.

Flesh Slightly creamy-white. Fine-textured. Juicy.

Aroma Nil.

Flowers Pollination group 3. Biennial.

Leaves Medium to rather small. Acute. Crenate or bluntly serrate. Medium thick. Slightly upward-folding. Flat. Light yellowish-green. Fairly downy.

GOLDEN NOBLE

Season October to December
Picking time Early October

It was Patrick Flanagan, the gardener to Sir Thomas Hare of Stowe Hall, Downham in Norfolk, who discovered this apple growing in an old orchard and introduced it to the Horticultural Society of London in 1820. It is a handsome fruit and one of the best cookers being acid with an extremely good fruity flavour. It is a garden and exhibition variety and not grown commercially. Trees are available from several specialist nurseries. They make upright-spreading, moderately vigorous trees which are partial tip-bearers. The cropping is moderate.

Size Large, 80 × 65mm (3⅛ × 2½″). Large on young trees only, otherwise medium.
Shape Round, sometimes rather flattened and slightly conical. Fairly symmetrical though occasionally slightly lop-sided. Barely any trace of ribs. Very regular.
Skin Light green (144C) to rather ochre-green (151A) nearest the sun, becoming golden yellow (7D). Usually clear green or gold with no flush but there can be an area of slightly striped or mottled grey pinkish-brown nearest the sun. Lenticels not very conspicuous as tiny pinkish-brown russet dots within a larger dot of whitish-green, or just whitish-green dots. Some slight scarf skin at the base especially noticeable around the edges of the russet that extends from the cavity. Skin smooth becoming greasy.
Stalk Stout (4mm) and short (9mm), sometimes fleshy. Set well down within cavity.
Cavity Rather narrow. Medium depth. Sometimes with a fleshy lip on one side.
Eye Medium to small. Slightly to half open. Sepals fairly erect and convergent, broad based with tips reflexed. Fairly downy.
Basin Rather shallow. Medium width. Slightly ribbed and puckered.
Tube Long funnel shaped.
Stamens Marginal.
Core line Faint, median.
Core Median. Axile, closed or open.
Cells Roundish obovate.
Seeds Acuminate or acute. Straight not curved. Mid brown, quite large.
Flesh Creamy-white. Slightly soft. Fairly fine-textured. Fairly juicy.
Aroma Very slight.
Flowers Pollination group 4.
Leaves Medium size. Acute. Bluntly pointed serrate. Medium thick. Slightly upward-folding, flat not undulating. Mid green. Undersides downy.

SUNSET

Season October to December
Picking time Late September

A high quality mid to late season dessert apple which was raised in England about 1918. It was raised by Mr G.C. Addy at Ightham in Kent from a pip of Cox's Orange Pippin and introduced jointly by Mr Addy and Mr William Rogers of Dartford in Kent. It was named in 1933. In 1960 it received the Award of Merit from the RHS and a First Class Certificate in 1982. It is a good garden variety, being too small for commercial use, and can be grown in areas where Cox's Orange Pippin may fail, being reasonably hardy and tolerant of a moist climate. The trees are upright-spreading yet compact in shape, and are of moderate vigour. They produce spurs very freely. The cropping is good and reliable though the fruits need to be thinned in order to achieve a decent size. Trees are widely available from nurseries.

Size Medium, 61 × 51mm (2⅜ × 2″).
Shape Flat round to round-conical. Symmetrical or a little lop-sided. Slight trace of well rounded ribs. Fairly regular.
Skin Greenish yellow (150B) becoming yellow (12B). Quarter to three quarters flushed with bright orange (169C) to orange-red (34A). Distinct broad broken stripes of crimson-red (46A). Frequently numerous small patches of fine greyed-ochre or grey-brown russet. Some lenticels fairly conspicuous as quite large grey-brown russet dots, which are sometimes angular or star-shaped. Skin smooth and dry, sometimes slightly textured.
Stalk Medium to stout (3–4.5mm). Medium to fairly long, (15–22mm). Extends well beyond base.
Cavity Wide. Medium to fairly deep. Regular. Partly or more lined with fine ochre-brown or scaly grey-brown russet which can scatter over base.
Eye Medium size. Half open. Sepals erect convergent with tips reflexed. Slightly downy.
Basin Medium depth. Medium to fairly wide. Regular or slightly ribbed. Ochre or grey russet runs concentrically around basin.
Tube Funnel-shaped.
Stamens Median.
Core line Median towards basal.
Core Median. Axile.
Cells Obovate or roundish.
Seeds Large. Obtuse. Broad, not very plump.
Flesh Creamy-white. Firm but tender. Slightly coarse-textured. Slightly juicy.
Aroma Slight, rather sharp aroma.
Flowers Pollination group 3.
Leaves Medium to small. Broadly oval. Bluntly and quite broadly serrate. Fairly thin. Slightly upward-folding. Flat. Mid green. Downy.

— GASCOYNE'S SCARLET —

Season   October to January
Picking time   Mid September

A very attractive mid to late season dual purpose apple which was raised by Mr Gascoyne of Bapchild Court, Sittingbourne, Kent. It was introduced in 1871 by G. Bunyard & Co. of Maidstone and received a First Class Certificate from the RHS in 1887. It is not grown commercially. The trees are very vigorous, and rather gaunt, upright-spreading in habit and partial tip-bearers. The cropping is moderate and the fruits fairly sweet with a refreshing acidity. The skins are tough. When cooked the flesh is pale greenish-yellow, sub acid with quite a good flavour. It breaks up completely but not to a fluff. Available on a small scale from specialist nurserymen. The trees are triploid.

Size Large, 77 × 64mm (3″ × 2½″).
Shape Flat-round. Flattened at base and apex. Five crowned at apex. Large fairly prominent well-rounded ribs. Symmetrical. Fairly regular.
Skin Pale yellow-green (145A) to pale primrose yellow (154D). Half to nearly completely covered with bright crimson-red flush (46A). Some indistinct short broken stripes of a deeper crimson (53A). Numerous lenticels, very pronounced, showing as green or greenish-ochre russet dots, some of which are quite large. Because of the number of green lenticels the unflushed skin appears rather greener from a distance. Skin slightly greasy and smooth.
Stalk Medium to fairly stout (3–3.5mm). Medium to fairly long (15–25mm). Stalk set deep into cavity so level with base or protruding slightly beyond.
Cavity Deep and fairly narrow. Regular cone-shape. Dark ochre green with variable scaly grey-brown russet, which can streak over base.
Eye Large. Partly open. Sepals broad-based, long and sharply pointed. Erect convergent with tips reflexed.
Basin Wide and very deep. Prominently ribbed and puckered. Fairly downy.
Tube Cone-shaped.
Stamens Median towards basal.
Core line Basal clasping, sometimes joining.
Core Median to somewhat distant. Axile or abaxile.
Cells Obovate.
Seeds Obtuse. Oval and fairly plump. Regular.
Flesh White tinged green, cream under the skin on the flushed side. Firm, fine-textured and juicy.
Aroma Sweetly aromatic.
Flowers Pollination group 5. Triploid.
Leaves Large. Broadly oval. Serrate. Medium to rather thin. Slightly undulating, sometimes downward-folding. Dark-green. Fairly downy.

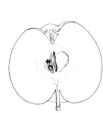

— RIBSTON PIPPIN —

Season   October to December
Picking time   Early October

This highly prized late dessert was discovered at Ribston Hall near Knaresborough in Yorkshire and it is thought to have been raised from seed brought there from Rouen in about 1688. The original tree was blown down in 1810 but supported by stakes, continued to bear fruit until it died in 1835. A young shoot sprouted from the stem and grew into a tree until that too was blown down in 1928. It received the Award of Merit from the RHS in 1962.

The trees are reasonably hardy, vigorous, upright-spreading and produce spurs very freely. They are triploid. The cropping is moderate to good but the fruits are liable to sudden drop at harvest time. The flavour of the fruit is good with a nice balance of sugar and acid, with a slight hint of pear-drops. The skin is a little chewy. The fruit becomes rather dry by December and is better if eaten earlier. Available from a few specialist nurserymen.

Size Medium large, 70 × 58mm (2¾ × 2¼″).
Shape Round-conical. Flattened at base and sometimes at apex. Frequently lop-sided. Some unequally large ribs makes the fruit irregular. Sometimes flat-sided. Slightly five-crowned at apex.
Skin Fairly strong greenish-yellow (151D). Quarter to three quarters flushed with brownish-orange (171B) with numerous broad broken stripes of red (45A). Rather scratchy patches of greenish-brown russet, mostly at base but some on cheeks, and often some grey-brown russet at apex. Lenticels reasonably conspicuous as greenish-brown russet dots and some pale grey dots towards base. Skin smooth and fairly dry, becoming greasy.
Stalk Medium thick (3mm) and fairly short (10mm) usually within cavity or level with base.
Cavity Narrow and deep. Lined with grey-brown scaly russet which can spread over base.
Eye Medium size to rather small. Closed or slightly open. Sepals erect connivent with tips reflexed. Very downy.
Basin Medium to deep and fairly wide. Ribbed and slightly puckered. Occasional trace of beading. Usually some fine cinnamon or grey-brown russet.
Tube Long cone or funnel-shaped.
Stamens Median or basal.
Core line Median inclined towards basal. Usually joining below stamens.
Core Median. Axile, sometimes slightly open.
Cells Obovate.
Seeds Few, long oval, acuminate. Fairly plump.
Aroma Almost nil, slightly vinous when cut.
Flowers Pollination group 2. Triploid.
Leaves Medium size. Broadly oval. Serrate. Medium thick, flat. Upward-folding. Mid green. Downy.

TOWER OF GLAMIS

Season   November to February
Picking time   Mid to end of October

A very old culinary apple of Scottish origin, known before 1800. The exact history of this apple appears to be lost but according to the *Herefordshire Pomona* of 1876–85 the variety 'abounds in the orchards of Clydesdale and Carse of Gowrie'. It is not grown commercially. It makes an upright-spreading rather compact tree of moderately vigorous growth, which is a spur-bearer. The trees are hardy and suitable for growing in the north. The fruit is sub-acid, rather tasteless but quite pleasant with the addition of some honey. It cooks to a fluff. Available from one or two specialist nurseries.

Size Medium large, 73 × 70mm (2⅞ × 2¾").
Shape Round-conical to oblong-conical. Large well-rounded or angular ribs, frequently slab-sided. Irregular. Can be four-sided with four prominent ribs instead of five. Distinctly four or five crowned at apex.
Skin Fairly bright grass green (144A is the nearest but too dark). Turning sulphur yellow (151A). Has a bloomed appearance due to much scarf skin that starts at base and covers about half the fruit sometimes extending to the apex. Lenticels indistinct whitish-green dots or brown russet dots. Skin smooth and very slightly greasy.
Stalk Medium width (3mm) and fairly short to medium (12–15mm). Level with base of fruit.
Cavity Fairly narrow and medium depth. Partly or completely russetted with golden-brown scaly russet which can streak and scatter over base.
Eye Medium size. Closed or partly open. Sepals broad-based, erect convergent. Very downy.
Basin Rather narrow and usually fairly deep. Much puckered and ribbed and rather pinched looking.
Tube Long cone or slightly funnel-shaped sometimes reaching almost into core cavity.
Stamens Median sometimes towards marginal.
Core line Median, but below stamens.
Core Median. Large and open. Abaxile.
Cells Obovate or roundish obovate. Can be rather elliptical. Tufted.
Seeds Acute. Round, pointed or bluntly pointed, plump. Short and small. Fairly regular.
Flesh White tinged slightly green. Firm and crisp. Coarse-textured. Juicy.
Aroma Slight, rather like pears. Stronger when cut.
Flowers Pollination group 4.
Leaves Large. Broadly oval. Serrate. Very thick and leathery. Slightly undulating. Dark green. Downward-hanging. Undersides very downy.

GALA

Season   October to early January
Picking time   Early October

This mid to late season dessert apple is a fairly new introduction to the UK. It was raised in New Zealand in 1934 by J.H. Kidd of Greytown, Wairarapa from Kidd's Orange Red × Golden Delicious. It was selected in 1939 and introduced in 1960. It was named in 1965. Commercially its popularity is open to speculation because of its fruit size and rather indefinite colour. The flavour is erratic: it can be very sweet and interesting with a hint of pear-drops, but it can also be rather flat in some seasons. It tends to fade with keeping. The trees are spreading in habit and of moderate vigour and they produce spurs freely. They are suitable for growing in the North but are prone to scab. The cropping is good to heavy. Fairly widely available from nurserymen.

Size Medium, 61 × 58mm (2⅜ × 2¼").
Shape Oblong-conical. Distinct fairly narrow ribs, pariculary noticeable at apex where they terminate in five rather flat crowns. Fairly symmetrical but can be slightly lop-sided. Fairly regular.
Skin Whitish-green (between 145B and 150C) becoming bright yellow (13B). Quarter to almost completely flushed with dense crimson red dots and flecks (46A) under which the skin is yellow or orange-red (31B). Short narrow broken stripes of crimson-red (46A). Lenticels fairly conspicuous greenish or grey-brown russet dots. Some small patches of ochre russet. Some thin purplish scarf skin at base. Skin smooth and dry.
Stalk Slender to fairly slender (2–2.5mm). Very long (25mm). Set at an angle in the cavity.
Cavity Wide and fairly deep, sometimes lipped. Partly lined with greenish-golden russet overlaid with fine grey scaling.
Eye Medium size. Closed or slightly open. Sepals very long, broad based, tapering to a sharp point, convergent with half to two thirds reflexed. Sepals green with tips brown. Slightly downy.
Basin Wide and deep. Pronounced ribs. Slightly puckered. Occasional small patches of grey russet.
Tube Funnel-shaped.
Stamens Median.
Core line Rather indefinite. Basal.
Core Median. Axile, sometimes slightly open.
Cells Ovate.
Seeds Obtuse. Quite large, fairly plump. Regular.
Flesh Yellowish-cream. Firm but slightly soft. Fine-textured. Not particularly juicy.
Aroma Sweetly aromatic.
Flowers Pollination group 4.
Leaves Medium to small. Narrow acute. Serrate. Rather thin. Flat not undulating. Upward-folding. Dark green. Undersides not very downy.

MARGIL

Season   October to January
Picking time   Early October

The history of this old dessert apple is uncertain. One theory is that it was brought to England by George London who worked in the gardens at Versailles under De La Quintinye, and who was a partner of the Brompton Park nursery where this apple was extensively cultivated as early as 1750. Rogers in his *Fruit Cultivator* of 1834 states that he has known it for seventy years, it being then in repute as a dessert fruit. The first tree he saw of it was an espalier planted by Sir William Temple in the Sheen Garden. Margil is still listed today by a few specialist nurserymen. They make small trees of weak growth, well suited to the small garden. They are hardy but susceptible to late frosts due to the early flowering. They are spur-bearers. The cropping is good and the fruits are sweet with a rich flavour.

Size Small, 54 × 51mm (2⅛ × 2″).

Shape Round-conical. Symmetrical or lop-sided. Prominent irregular ribs often making it angular and flat-sided. Irregular. Five-crowned.

Skin Greenish-yellow (151D) to yellow (12B). Half to three quarters flushed with orange-red (169A) to crimson-red (46A) on well coloured fruits. Distinct fairly broad, broken or long stripes of crimson-red (46A) which are paler red when over the yellow skin. Some fine ochre-grey or greenish-grey russet paches. Some lenticels conspicuous as grey-brown or ochre-grey russet dots. Skin smooth and slightly greasy.

Stalk Slender (2mm) medium length (13–18mm). Protrudes well beyond base.

Cavity Narrow and quite deep with stalk deeply inserted. Sometimes lipped. Lined with greenish-grey scaly brown russet which radiates out.

Eye Fairly small. Closed or partly open. Sepals broad-based, erect and rather pinched together.

Basin Rather narrow. Medium depth. Distinctly ribbed, sometimes beaded. Often falls away on one side. Usually some fine grey-brown russet.

Tube Deep, cone or slightly funnel-shaped.

Stamens Median.

Core line Median sometimes inclined towards basal.

Core Median. Axile.

Cells Roundish ovate or round.

Seeds Acuminate or acute. Quite large, oval. Sharply or bluntly pointed. Straight or curved.

Flesh Creamy-white tinged slightly green. Firm. Slightly coarse-textured. Rather dry.

Aroma Sweetly aromatic.

Flowers Pollination group 2.

Leaves Medium to rather small. Acute to narrow acute. Crenate. Medium thick. Upward-folding. Flat. Mid grey-green. Moderately downy.

SUNTAN

Season   November to January
Picking time   Mid October

A new late dessert apple raised in England in 1955 by Dr Alston of East Malling Research Station in Maidstone, Kent, from Cox's Orange Pippin × Court Pendu Plat. It is a good late dessert apple, later than Cox's Orange Pippin, with a very acid Cox-like flavour. Because of its late flowering it is a useful variety to grow where frost is a problem. It is a triploid but, in difficult areas with suitable pollinators and good management, it could produce a crop where other varieties may fail. The trees are very vigorous and for the small garden or restricted form should be grafted on to a dwarfing rootstock. The habit is wide-spreading and they produce spurs freely. The cropping is good. Trees are widely available.

Size Medium large, 70 × 54mm (2¾ × 2⅛″).

Shape Flat-round. Distinctly flattened at base and flattened at apex. Slight trace of ribs. Symmetrical or slightly lop-sided. Fairly regular.

Skin Rather dull greenish-yellow (4A). Quarter to three-quarters flushed dull orange-red (34A). Rather indistinct short, fairly broad, broken stripes of dull crimson-red (46A). Lenticels very conspicuous quite large ochre-grey or grey-brown russet dots. Patches of ochre-grey or pale grey-brown russet making surface textured, otherwise smooth. Dry becoming slightly greasy.

Stalk Fairly stout to stout (3.5–4.5mm). Medium length (15–20mm) protruding beyond base, or short (8mm) and level with base.

Cavity Medium width and depth. Regular. Lined with greenish-ochre overlaid with greyish-brown slightly scaly russet which streaks out over base.

Eye Large, half open. Sepals broad based, erect or rather flattish convergent with tips reflexed or broken off. Fairly downy.

Basin Wide and shallow. Slightly puckered with some beading. Sometimes lined with fine pale grey-brown russet which can extend over apex.

Tube Short, broad funnel-shaped.

Stamens Median.

Core line Rather faint. Median joining at base of funnel bowl.

Core Somewhat sessile. Axile.

Cells Obovate.

Seeds Large. Obtuse. Quite plump. Regular.

Flesh Yellowish-cream. Firm and fairly juicy. Rather coarse-textured. Crisp.

Aroma Almost nil, very slightly woody.

Flowers Pollination group 5. Triploid.

Leaves Quite large. Rather long acute. Bluntly serrate. Medium thick. Flat not undulating. Upward-folding. Mid greyish-green. Very downy.

KIDD'S ORANGE RED

Season   November to January
Picking time   Early October

A mid to very late dessert apple, introduced to England about 1932 from New Zealand where it was raised in 1924 by J.H. Kidd of Greytown, Wairarapa from Cox's Orange Pippin × Delicious. It is one of the finest flavoured dessert apples, sweet and aromatic, but fruits can be small unless thinned. It is grown commercially in New Zealand and gained some popularity here because it was thought that it might supplement Cox's Orange Pippin as it showed promise of storing better. It is however prone to excessive russetting in some districts which detracts from its appearance and it is susceptible to damage by certain sprays. The trees are moderately vigorous, upright-spreading and produce spurs quite freely. The cropping is good. Trees are available from several specialist nurseries. There is a highly coloured good flavoured sport available called Captain Kidd.

Size Medium, 67 × 64mm (2⅝ × 2½").
Shape Conical. Rounded at shoulder, but flattened at base. Narrow at apex. Fairly distinct ribs and sometimes flat-sided towards apex. Slightly five crowned at apex. Fairly regular.
Skin Pale greenish-yellow (154C) becoming pale yellow (8B). Half to almost completely covered with crimson flush (53B) which is slightly more red than crimson at the edges (46A). Indistinct narrow broken stripes of purplish-crimson (187B). Some small patches of pale grey-brown or greenish-ochre russet. Some indistinct scarf skin at base. Lenticels not very conspicuous grey-brown russet dots. Skin smooth and dry.
Stalk Fairly stout to stout (3.5–4mm). Medium to fairly long (15–20mm). Protrudes just or well beyond base. Often fleshy at spur attachment.
Cavity Wide and deep. Partly or completely lined with golden-grey russet which can streak over base and scatter over cheeks.
Eye Small to medium. Closed. Sepals erect convergent with tips reflexed. Very downy.
Basin Medium width and shallow. Ribbed. Some fine brown russet present. Sometimes downy.
Tube Long cone-shaped or funnel-shaped.
Stamens Median.
Core line Rather faint. Basal, clasping.
Core Median. Axile or abaxile.
Cells Round to roundish-ovate.
Seeds Obtuse to acute. Large. Numerous. Regular.
Flesh Creamy-white. Firm and juicy. Fine-textured.
Aroma Very slightly vinous.
Flowers Pollination group 3.
Leaves Medium to small. Acute. Bluntly serrate. Medium thick. Flat not undulating. Slightly upward-folding. Mid greyish-green. Fairly downy.

BLENHEIM ORANGE

Season   November to January
Picking time   Early October

The Blenheim Orange is one of the loveliest apples of all with its dry distinctive flavour. It was found at Woodstock near Blenheim in Oxfordshire in about 1740. It is recorded that a countryman named Kempster planted the original kernel and the apple, known locally as Kempster's Pippin began to be catalogued in about 1818. It received the Banksian Silver medal in 1820 and thereafter spread through England to Europe and America. The distinctive Classical Blenheim has descended from Kempster's original tree but many clones exist and it seems to be the Broad-Eyed Blenheim that is most common. It is a dual-purpose triploid apple with vigorous growth that requires a dwarfing rootstock for restricted space or form. It is a partial tip-bearer and fairly resistant to mildew. The trees bear shyly when young but improve with age. They are widely available from nurserymen.

Size Large, 83 × 70mm (3¼ × 2¾").
Shape Flat-round. Flattened base and apex. Often lop-sided. Faint well-rounded ribs. Slightly five-crowned at apex. Fairly regular.
Skin Rather dull yellowish-green (between 144A & 144C), becoming yellow (7A). Slightly to half flushed with speckled dull orange (169B) to orange-red (34B) becoming orange-yellow (167B) at the edges. Indistinct short broad stripes of crimson-red (46A). Speckled and dotted with fine grey-brown russet. Lenticels grey-brown or pale ochre russet dots. Skin smooth and dry becoming greasy.
Stalk Stout (4mm) and fairly short (10–14mm) within cavity or protruding very slightly beyond.
Cavity Fairly deep and quite wide. Sometimes lipped. Lined with greenish-ochre and fine slightly scaly light brown russet which usually straggles over base.
Eye Large. Wide open. Sepals well separated at base, erect with tips usually reflexed or broken off. Not very downy.
Basin Wide. Quite deep. Ribbed. Slightly russeted.
Tube Wide slightly funnel-shaped.
Stamens Median.
Core line Basal, clasping.
Core Median. Axile, open.
Cells Obovate.
Seeds Acuminate. Long, thin and pointed.
Flesh Creamy-white. Slightly coarse-textured. Firm but tender and crisp.
Aroma Slightly vinous aroma.
Flowers Pollination group 3. Triploid.
Leaves Fairly large. Broadly ovate or broadly acute. Finely and sharply serrate. Medium thick. Flat. Dark green. Slightly downward-hanging. Downy.

CHIVER'S DELIGHT

Season   November to January
Picking time   Mid October

An attractive late dessert apple which was raised in England about 1920 by Mr Chivers of Chivers Farms, Histon, Cambridgeshire. The parentage is unknown. There are a few commercial plantings in the United Kingdom. The trees are moderately vigorous, upright in habit and produce spurs very freely. The trees are fairly hardy and suitable for growing in the North. The fruits are sweet with a slightly acid bite and with an interesting flavour, but they are erractic and sometimes quite flat-tasting. The cropping is good and the fruit keeps well. This apple is listed by several specialist nurseries.

Size Medium, 68 × 58mm (2⅝ × 2¼").
Shape Flat-round to round. Base can be wider than apex. Flattened at base and apex. Some fruits lopsided. Slightly ribbed. Slightly five crowned at apex. Almost regular.
Skin Variable. Yellow-green (144C) becoming pale golden yellow (13B). Slightly to half dotted and mottled with red (45A) with the underlying orange-yellow (23A) skin showing through. A few short, broken, rather thin stripes of crimson-red (46A). Some well coloured fruits have a denser flush of crimson-red (46A) with little flecking and short thin broken stripes of purplish-carmine (59A). There is usually a fairly well defined demarcation line between flush and yellow skin. Some small scattered cinnamon-brown russet patches. Lenticels large pale grey-brown sometimes star-shaped spots.
Stalk Slender to medium (2–3mm). Very long (28–35mm). Stalk is thicker where it joins the fruit.
Cavity Wide to very wide. Medium depth. Partly or more lined with fine grey or scaly grey-brown russet which runs concentrically round cavity and can streak out over base. Can be lipped.
Eye Rather small. Closed or slightly open. Sepals narrow, tapering, erect with tips slightly reflexed or broken off. Stamens visible. Very downy.
Basin Medium depth and narrow. Ribbed and a little beaded. Some specks of cinnamon russet.
Tube Small. Long cone or slightly funnel-shaped.
Stamens Median or marginal.
Core line Basal, clasping.
Core Median. Axile.
Cells Round or obovate.
Seeds Acute. Broad. Frequently flattened on one side.
Flesh Creamy-white. Firm. Fine textured. Juicy.
Aroma Slightly sweetly aromatic.
Flowers Pollination group 4.
Leaves Medium size. Acute. Small, bluntly serrate. Rather thin. Flat. Some upward-folding. Very downy.

LADY HENNIKER

Season   November to January
Picking time   Mid September

This is an old English dual purpose apple raised between 1840 and 1850 by John Perkins, gardener at Thornham Hall, Eye in Suffolk, the seat of Lord Henniker. The seeds from the apples used to make cider were sown and the most promising seedlings were then selected and grown on. The tree in question was considered a favourite and was carefully preserved. It was finally introduced in 1873 when it received a First Class Certificate from the RHS. The flavour as a dessert is mildly acid and aromatic. When cooked the flesh becomes yellow, breaks up completely but not to fluff and remains acidic with a lovely flavour. The cropping is moderate to good. The trees are vigorous, upright-spreading and, produce spurs freely. The trees are fairly hardy. They have a very limited availability from nurseries.

Size Large, 73 × 70mm (2⅞ × 2¾").
Shape Oblong. Frequently lop-sided. Prominent ribs often with one larger than the rest. Frequently flat or slab-sided, with angular ribs, and concave between. Irregular. Distinctly five-crowned at apex.
Skin Bright, quite deep, yellow-green (144A) becoming yellow (12A). A trace to half flushed with thin ochre-orange (163B). Spotted and indistinctly striped with rather thin orange-red (169A). Can have some patchy scarf skin. Variable amounts of fine netted grey-brown russet. Lenticels dark grey-brown or whitish russet dots.
Stalk Fairly stout (3.5mm). Short or very short (8–15mm). Within cavity.
Cavity Deep and wide with stalk well sunk within. Can have one or two pronounced ribs. Some pale brown slightly scaly russet, often underlaid with dark green, which can streak out.
Eye Fairly large. Closed or slightly open. Sepals broad based, sometimes slightly separated at base, convergent with tips reflexed. Fairly downy.
Basin Very deep, rather narrow. Pronounced ribs. Traces of fine grey-brown or cinnamon russet.
Tube Variable. Large cone or funnel-shaped.
Stamens Median.
Core line Towards basal, clasping, sometimes appearing to pull out the sides of the tube.
Core Slightly distant. Abaxile.
Cells Obovate. Tufted.
Seeds Acute. Rather thin. Straight or curved.
Flesh Creamy-white. Firm. Rather coarse-textured.
Aroma Sweet and spicy.
Flowers Pollination group 4.
Leaves Fairly large. Broadly oval or broadly acute. Sharply serrate. Medium thick. Flat or slightly undulating. Mid green. Downward-hanging. Very downy.

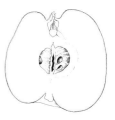

ORLEANS REINETTE

Season   November to January
Picking time   Mid October.

This late dessert apple is thought to have originated in France. Edward Bunyard recalls in 1920 that the variety was 'first described by Knoop in 1776' but does not elucidate. In 1914 it received the Award of Merit from the RHS as Winter Ribston; and again in 1921, as Orleans Reinette. It then came to general notice. It is widely listed today. The fruit has a lovely flavour with a suggestion first of sweet oranges, followed by a nutty flavour after the initial juiciness has gone. However they are subject to sudden drop at harvest time and have to be watched. They are best stored in polythene bags to avoid shrivelling, and the cropping is rather irregular. The trees are vigorous, upright-spreading and spur-bearers.

Size Medium large, 73 × 61mm (2⅞ × 2⅜″)

Shape Flat-round. Distinctly flattened at base and apex. Symmetrical or slightly lop-sided. Trace of well rounded ribs. Fairly regular.

Skin Yellowish-green (144C) becoming dull golden yellow (12B). Slight to three quarters flushed with dull orange (171B) to orange-red (34B) with a few rather indistinct short, broken stripes of scarlet (45B). A great deal of the surface is covered with flecks, jagged patches and areas of fine, slightly scaly greyed yellow (161A) greyed brown (199C) russet. Lenticels conspicuous as jagged and sometimes angular russet dots. Skin dry and textured.

Stalk Fairly stout to stout (3.5–4mm). Short to medium (10–20mm). Within cavity or beyond.

Cavity Deep and fairly wide. Sometimes lipped. Green, and completely lined with fine grey-brown russet, which comes out over base.

Eye Very large. Wide open with stamens clearly visible. Sepals separated at base, flat convergent with tips well reflexed or broken off.

Basin Wide. Medium depth. Very slightly ribbed but regular. Usually some cinnamon russet which appears to run concentrically round basin.

Tube Short, wide, cone-shaped.

Stamens Median.

Core line Basal, fairly prominent, clasping.

Core Sessile. Axile, sometimes open.

Cells Obovate or roundish.

Seeds Acute. Fairly plump, sometimes angular.

Flesh Creamy white. Firm. Fine-textured. Very juicy.

Aroma Nil.

Flowers Pollination group 4. Can be biennial.

Leaves Quite large. Broadly oval. Sharply serrate. Fairly thin. Flattish. Very slightly upward-folding. Dark green. Undersides not downy.

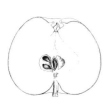

JUPITER

Season   November to January
Picking time   Early October

This is a fairly recent dessert apple raised by East Malling Research Station in Kent in 1966 from Cox's Orange Pippin × Starking. It was named in 1973. It has been planted up commercially here but may be superseded by more recent introductions such as Fiesta. It is a late season apple, juicy and sweet with a full aromatic flavour. The skin is a little chewy. The trees are fairly hardy and are suitable for growing in the North. They are vigorous, spreading in habit and spur-bearers and they are triploid. This apple is listed by a few fruit tree nurserymen.

Size Medium, 64 × 58mm (2½ × 2¼″).

Shape Conical. Flattened at base and apex. Sometimes symmetrical, sometimes a little lop-sided. There can be some slight well rounded ribs. Slightly five-crowned at apex. Mostly regular.

Skin Greenish-yellow (144C) becoming rather dull golden yellow (151A). Half to three quarters flushed with orange-red (34B). Very distinct long broad stripes of crimson-red (46A). Stripes often present over yellow skin where they are fainter. A good deal of scarf skin at the base and sometimes on the cheeks often giving the fruit a milky-lilac and mottled appearance. Lenticels fairly noticeable as whitish dots often surrounded by an areola of pink colour. Skin smooth and dry.

Stalk Medium to stout (3–4mm) and long (25mm). Frequently more stout where it joins the fruit. Often set at an angle.

Cavity Wide. Medium to rather shallow. Nearly always lipped on one side. Usually lined with greenish-ochre or grey-brown finely scaled russet which can streak slightly over base.

Eye Medium size. Partly or fully open. Sepals long, broad based, erect convergent with tips reflexed. Fairly downy.

Basin Fairly wide and deep. Ribbed. Sometimes a small amount of light brown russet.

Tube Funnel-shaped.

Stamens Median.

Core line Median, joining at base of funnel bowl.

Core Median. Axile.

Cells Obovate.

Seeds Acuminate. Rather shrivelled and starved looking.

Flesh Creamy-white. Rather coarse-textured. Juicy.

Aroma Quite strong, sweetly scented.

Flowers Pollination group 3. Triploid.

Leaves Small. Acute. Serrate. Thick and leathery. Upward-folding and slightly undulating. Mid grey-green. Undersides downy.

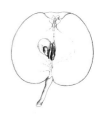

LAXTON'S SUPERB

Season November to January
Picking time Early October.

This late dessert apple was raised in England in 1897 by Laxton Bros Ltd. of Bedford from Wyken Pippin × Cox's Orange Pippin. It received an Award of Merit from the RHS in 1919 and a First Class Certificate in 1921. It was introduced in 1922 and is grown commercially today. When introduced it was widely planted as the young trees cropped regularly and heavily, but on becoming established they tend to become biennial. The fruit can vary considerably in flavour from season to season and area to area and it can have an almost Cox-like flavour in hot summers. The trees are fairly hardy and are suitable for growing in the North. Trees are vigorous, upright-spreading and spur-bearers, but are prone to scab. Quite widely listed by nurserymen.

Size Medium to medium large, 67 × 61mm (2⅝ × 2⅜″) or 70 × 61mm (2¾ × 2⅜″).

Shape Round-conical to conical. Indistinct well rounded ribs. Symmetrical or slightly lop-sided. Fairly regular.

Skin Pale greenish-yellow (1B). Half to almost completely covered with dull purplish-red (185A), less dense and rather mottled away from the sun. A few short broken stripes of dark purplish-crimson. Some small dirty greenish-ochre russet patches. Lenticels fairly conspicuous quite large grey-brown dots which are sometimes star-shaped. There can be some fine, slightly broken scarf skin. Skin dry and slightly rough. Superb is often confused with Allington Pippin.

Stalk Medium thick (3mm) and medium to fairly long (15–20mm). Protrudes well beyond base.

Cavity Medium width and depth. Regular. Lined with greenish-ochre over which there is some scaly dark grey-brown russet which can come out over the shoulder. Sometimes lipped on one side.

Eye Quite large. Open or partly open. Sepals fairly broad based, erect with tips reflexed or broken off.

Basin Medium width and depth. Regular very slightly puckered.

Tube Cone shaped.

Stamens Median.

Core line Basal, clasping.

Core Median. Axile.

Cells Ovate, sometimes roundish.

Seeds Acuminate. Fairly plump. Regular.

Flesh White tinged slightly green. Fine-textured. Firm but tender. Juicy. Skin a little tough.

Aroma Nil.

Flowers Pollination group 4. Biennial.

Leaves Medium size. Acute. Sharply or bluntly serrate, sometimes broadly serrate. Fairly thin. Flat. Upward-folding. Mid greyish-green. Downy.

COX'S ORANGE PIPPIN

Season Late October to January
Picking time Early to mid October

Cox's Orange Pippin is regarded as the finest of all English apples and is the most extensively planted dessert variety in the UK. It was raised from pips of a Ribston Pippin in about 1825 at Colnbrook Lawn, near Slough, Bucks where the original tree grew until destroyed by a storm in 1911. The man who raised this historic fruit was Richard Cox (1777–1845) a retired brewer from Bermondsey. It was introduced by Charles Turner in about 1850 and received the Award of Merit and a First Class Certificate from the RHS in 1962. The trees are moderately vigorous, upright-spreading and spur-bearers. They are prone to canker in cold, wet areas and susceptible to scab and mildew. It is not suitable for the North preferring a warmer climate and it needs good soil conditions and a favourable environment to crop well. This apple is widely available from fruit tree nurserymen.

Size Medium, 64 × 54mm (2½ × 2⅛″).

Shape Round-conical. Very slight trace of well rounded ribs. Usually symmetrical. Slightly flattened at base and apex. Regular.

Skin Yellowish-green (144C) becoming clear yellow (9A–12A). Quarter to three quarters flushed with orange-red (34A) to greyed-red (178B) on more coloured fruits. Some broken stripes of brownish-crimson (185A). Some patches and dots of fine grey-brown russet. Some patchy scarf skin mostly towards base, giving the fruit a rather mottled appearance. The fruit brightens in colour as it matures. Skin dry and fairly smooth.

Stalk Fairly slender to medium (2.5–3mm). Medium length (15–20mm). Extends beyond base.

Cavity Fairly wide and medium depth. Often lipped. Lined with ochre-green and scaly grey-brown russet which usually extends over base.

Eye Fairly small. Half open. Sepals rather narrow, erect convergent, and three quarters reflexed. Fairly downy. Stamens often present.

Basin Rather shallow. Quite wide. Slightly ribbed. Variable amounts of grey-brown russet.

Tube Funnel-shaped.

Stamens Median.

Core line Basal, clasping.

Core Median. Axile.

Cells Obovate.

Seeds Obtuse. Quite large for size of apple. Wide.

Flesh Cream. Firm. Fine-textured. Juicy.

Aroma Slightly aromatic.

Flowers Pollination group 3.

Leaves Medium size. Acute. Bluntly serrate. Medium to rather thin. Slightly upward-folding. Mid yellowish-green. Undersides downy.

ADAM'S PEARMAIN

Season   November to March
Picking time   Early to mid October

Mr Robert Adams first brought this old English dessert apple to notice in 1826 under the name of Norfolk Pippin. Robert Hogg states in *The Fruit Manual* that it was exhibited in Herefordshire as Hanging Pearmain, and that it originated in that county.

The trees are fairly hardy and show some resistance to scab, making them suitable for growing in the West. The trees are moderately vigorous, widespreading and partial tip-bearers and have a biennial tendency. The cropping is good and the high quality fruits have the typical rather dry, nutty flavour of many of the russets. It is available from a few specialist nurseries.

Size Medium, 61 × 64, 70 or 73mm. (2⅜ × 2½, 2¾ × 2⅞″).
Shape Conical to long-conical. Sometimes waisted. Broad and rounded at base tapering to a flattened apex. Sometimes symmetrical but frequently a little lop-sided. Barely any trace of ribs. Regular.
Skin Dull pale-green (145B) becoming pale golden-yellow (13C). One third to almost completely covered with either dull crimson-red (46A) or a brighter orangey-red (34A). Broken, rather inconspicuous stripes of crimson-red (46A). Lenticels numerous, either greenish-grey or whitish russet dots. Many smaller pale ochre dots towards apex. Variably covered with fine ochre-brown or grey-brown russet. Skin dry and slightly rough.
Stalk Medium to stout (3–4mm). Shortish to quite long (13–22mm). Level with base or protrudes.
Cavity Medium width and depth. Sometimes lipped on one side. Lined with greenish-ochre or grey-brown russet which spreads out over base.
Eye Medium size. Slightly to half open. Sepals quite broad, sometimes separated at base, erect convergent with tips reflexed. Slightly downy.
Basin Fairly shallow. Medium width. Regular. Slightly ribbed and puckered. Sometimes slightly beaded. Occasional areas of ochre-brown russet.
Tube Funnel or cone-shaped.
Stamens Median.
Core line Basal or towards median.
Core Median. Abaxile.
Cells Variable. Usually obovate, sometimes ovate.
Seeds Acuminate. Plump. Regular. Mid brown.
Flesh Creamy-white. Firm. Fine-textured. Crisp but tender. Dry.
Aroma Nil.
Flowers Pollination group 2. Biennial.
Leaves Small. Oval. Serrate. Rather thin. Flat. Slightly downward-folding. Mid grey-green. Downy.

BISMARK

Season   November to February
Picking time   Late September

The origin of this mid to late culinary apple is uncertain. It is thought by some that it may have come from Bismark, Tasmania, around the 1870s, but others suggest that Victoria in Australia or Canterbury in New Zealand may have been the place of origin. It was named after Prince Bismark, the German Chancellor. In 1897 it received a First Class Certificate from the RHS. It is not grown commercially in the UK but is listed on a limited scale by fruit tree nurserymen. The trees are very hardy and suitable for growing in the North. They are moderately vigorous, spreading in habit and partial tip-bearers. The cropping is good to heavy. The flavour is sub-acid, strong and fruity and the fruit cooks to a greenish-yellow fluff.

Size Large, 77 × 70mm (3 × 2¾″).
Shape Round-conical. Flattened at base tapering to a narrow flat apex. Five-crowned at apex. Frequently a little lop-sided. Fairly distinct ribs and can be a little flat-sided. Fairly regular.
Skin Yellowish-green (145A) becoming pale slightly greenish-yellow (150C). Half to three quarters flushed with brilliant red (46B) deepening to crimson (46A) on sunny side with broken stripes of carmine (53A). There is sometimes a distinct patch of green or yellow on the flush where the skin has been shaded by a leaf. Lenticels either fairly conspicuous grey-brown russet dots surrounded by a circle of carmine, or tiny grey or whitish dots. Skin smooth and dry.
Stalk Fairly stout to stout (3.5–4mm). Short to medium length (10–18mm).
Cavity Moderately wide and deep. Cavity greenish and variably lined with finely scaled darkish brown russet which can radiate out over base.
Eye Medium size. Closed. Sepals erect convergent and rather pinched in. Occasionally the eye can be slightly open. Fairly downy.
Basin Deep and rather narrow. Sometimes slightly rectangular. Ribbed and wrinkled.
Tube Broad cone-shaped to slightly funnel-shaped.
Stamens Median.
Core line Median or basal, clasping.
Core Median. Axile.
Cells Round to roundish-obovate. Slightly tufted.
Seeds Sparse. Obtuse to acute. Quite large, plump.
Flesh White tinged green. Fine-textured. Crisp and juicy.
Aroma Nil, slightly acid when cut.
Flowers Pollination group 2. Biennial.
Leaves Medium size. Acute to broadly acute. Bluntly pointed serrate to crenate. Rather thin. Flat. Mid green. Slightly downward-hanging. Quite downy.

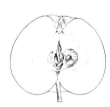

GOLDEN DELICIOUS

Season   November to February
Picking time   Late October

Golden Delicious is the most widely planted apple in the main fruit growing areas of the world. It is not always a great success, however, as it requires a reasonably high temperature and a greater light intensity than is usually found, for example, in England. The fruit grown in cold areas may be small and unattractively russetted. This apple originated, not in France, as one might suppose from the glut of examples sent here from that country, but in America. It was produced as a chance seedling found by A.H. Mullins of Clay Country, West Virginia in 1890. The parentage is not certain but is thought possible to have been from Grimes Golden open pollinated. It was introduced by Stark Brothers in 1914. The trees are moderately vigorous, spreading in habit and produce spurs very freely. The cropping is good and regular. The flavour of the fruit is sweet, with a nice balance of acidity, making it refreshing and clear tasting, but rather weak. Trees are widely available from nurserymen.

Size Medium, 67 × 64mm (2⅝ × 2½″).
Shape Round-conical to oblong. Distinctly ribbed with rather angular ribs becoming more pronounced towards apex. Five-crowned at apex. Rounded at base. Symmetrical. Fairly regular.
Skin Pale greenish-yellow (7C) becoming yellow (12A). Can be flushed with pale orange (22A). No stripes. Lenticels fairly conspicuous as greenish-brown russet dots larger and more pronounced towards base. There can be some small grey-brown russet patches. Skin smooth and dry.
Stalk Slender (2mm) and very long (30–41mm). Protrudes well beyond base.
Cavity Deep. Rather narrow. Partly or more lined with fine grey-brown russet which can radiate.
Eye Medium size. Closed or slightly open. Sepals long and tapering to a fine point. Erect convergent with tips partly reflexed. Slightly downy.
Basin Medium to deep. Medium width. Distinctly ribbed. Occasional fine cinnamon-brown russet.
Tube Cone-shaped.
Stamens Median.
Core line Basal, clasping.
Core Median. Abaxile.
Cells Obovate, quite long and narrow, or elliptical.
Seeds Acute. Fairly plump. Fairly numerous.
Flesh Cream tinged green on less golden fruits. Crisp and juicy. Fine-textured.
Aroma Sweetly aromatic.
Flowers Pollination group 4. Fairly frost resistant.
Leaves Medium size. Long narrow acute. Serrate. Thin. Flat not undulating. Upward-folding. Mid yellowish-green. Undersides slightly downy.

BRAMLEY'S SEEDLING

Season   November to March
Picking time   Mid October

The Bramley's Seedling is the most popular culinary apple in the United Kingdom, occupying an acreage greater than all the other culinary apples combined. It was raised in a cottage garden in Church Street, Southwell, Nottinghamshire by Betsy Brailsford from a seed of unknown origin between 1809 and 1813. The original tree is still in excellent condition.

The apple was introduced by Merryweather, was first exhibited in 1876 and received a First Class Certificate from the RHS in 1883. The trees are triploid, very vigorous and spreading, and they are partial tip-bearers. They are fairly hardy but are susceptible to spring frosts. The cropping is heavy but can be biennial. They are acid yet sweet with a good flavour and plenty of juice. They cook to a pale cream fluff. Trees are widely available from nurserymen. There is a red sport named Crimson Bramley.

Size Large to very large, 89 × 64mm (3½ × 2½″).
Shape Flat-round. Flattened at base and apex. Five-crowned at apex. Irregular large ribs, sometimes angular, occasionally flat-sided or concave between ribs. Irregular. Frequently lop-sided.
Skin Bright green (144A) becoming pale greenish-yellow (151D). Can be quarter to three quarters flushed with thin greenish-ochre (153B) with some broad broken stripes and dots of greyed-red (179A). Lenticels conspicuous dark grey-brown or green dots. Some scarf skin at base. Skin smooth and shiny, becomes greasy on maturing.
Stalk Stout (4.5mm) and short (6–10mm). Within cavity or level with base.
Cavity Medium to fairly deep. Generally fairly wide. Partly lined with very streaky light grey-brown russet which can radiate out over shoulder.
Eye Large. Closed or partly open. Sepals broad based. Flat convergent, tips reflexed. Very downy.
Basin Wide. Medium depth. Ribbed and sometimes beaded. Some minute flecks of russet.
Tube Slightly funnel-shaped, almost a cone.
Stamens Median.
Core line Median, sometimes towards basal.
Core Median. Axile, open.
Cells Variable. Generally round. Sometimes slightly obovate or elliptical. Tufted.
Seeds Acuminate. Rather starved looking.
Flesh Yellowish-white tinged slightly green. Firm. Coarse-textured. Juicy.
Aroma Slightly acid.
Flowers Pollination group 3. Triploid.
Leaves Large. Broadly acute. Finely serrate. Thick and leathery. Slightly undulating and slightly upward-folding. Dark green. Very downy.

Season   November to February
Picking time   Late September

Season   November to February
Picking time   Late October

The origin of this old culinary apple is not known. The nomenclature is somewhat confused and it is not the Mère de Ménage of France. It was known in Europe in the late 1700s but it is not known exactly when it was introduced into this country. The synonyms, Lord Combermere and Combermere lead to the inference that perhaps someone by this name introduced it. The fruit is acid and becomes a rather dull colour when cooked and breaks up completely but not to a fluff. The flavour is pleasant. The trees are vigorous, upright-spreading and partial tip-bearers, and they crop well. It is not grown commercially in the UK and is only listed by one or two nurseries.

This is a late dessert apple which was raised by H.M. Tydeman at East Malling Research Station in Kent in 1949 from Cox's Orange Pippin × Jonathan. It was named in 1974. The fruit is pleasant though rather watery with a good balance of sugar and acid. It has good keeping properties. The trees are moderately vigorous, upright-spreading spur-bearers and produce spurs freely. The cropping is heavy. Trees are available from several specialist nurseries.

Size Large,   77 × 67mm   (3 × 2⅝″) to very large   86 × 67mm (3⅜ × 2⅝″).

Shape Flat-round, some fruits oblong. Distinct ribs sometimes with one larger. Very irregular. Frequently lop-sided. Distinctly five-crowned at apex sometimes with smaller crowns in between.

Skin Pale yellow-green (144C) to pale yellow (154D). Quarter to almost covered with flush varying from dark brownish-crimson (185A) to reddish-brown (175C) on shaded side. Fairly broad broken stripes of dark purplish-crimson (187B). Lenticels conspicuous fairly large pale green or whitish dots usually surrounded by a circle of green. Some scarf skin at base. Skin smooth and dry.

Stalk Stout to very stout (4–5mm). Short (10–14mm). Usually within cavity.

Cavity Deep. Variable width. Completely lined with greenish-ochre overlaid with some fine scaly cinnamon russet which streaks on to base.

Eye Large. Partly or fully open. Sepals broad based, erect, or flat convergent. Some tips reflexed. Sepals can be separated at base. Very downy.

Basin Medium depth to rather shallow. Fairly wide. Distinctly ribbed. Sides slightly downy.

Tube Broad and deep cone-shaped.

Stamens Basal to median.

Core line Almost basal.

Core Median. Abaxile.

Cells Roundish obovate. Tufted.

Seeds Acute. Fairly long, rounded or bluntly pointed. Rather angular. Fairly plump.

Flesh Greenish-white. Juicy. Firm and crisp. Rather coarse-textured.

Aroma Very slight.

Flowers Pollination group 3.

Leaves Fairly large. Broadly acute. Serrate. Medium thick. Flat or slightly undulating. Dark green. Undersides quite downy.

Size Medium,   67 × 67 mm   (2⅝ × 2⅝″)   or   64 × 58mm (2½ × 2¼″).

Shape Round-conical to conical. Usually symmetrical but can be slightly lop-sided. Rounded base tapering to narrow rather flattened apex. Usually no ribs or very slight trace. Regular.

Skin Pale yellowish-green (150B) to pale greenish-yellow (5C). Quarter to three quarters flushed with fairly intense crimson-red (46A). On less well coloured fruits the flush is more mottled and striped crimson-red (46A) with the underlying yellow skin showing through. Indistinct broad broken stripes of crimson-red (46A) sometimes over the yellow skin. Stripes frequently enter cavity. Some patchy scarf skin at base. Some small patches of fine greenish-ochre russet. Lenticels inconspicuous greenish-ochre or grey-brown russet dots. Skin smooth and dry.

Stalk Medium to fairly stout (3–3.5mm). Fairly long (18–22mm). Extends beyond base.

Cavity Medium to deep. Medium width. Cone-shaped. Sometimes lipped. Lined with fine greyed-ochre russet which can streak over base.

Eye Medium size. Slightly open. Sepals very long and tapering erect convergent with tips two thirds reflexed. Very downy.

Basin Wide. Fairly deep. Quite distinctly ribbed. There can be some fine brown russet.

Tube Cone-shaped or slightly funnel-shaped, with fleshy protrusion at base entering the tube.

Stamens Median.

Core line Basal, clasping.

Core Median. Axile.

Cells Ovate.

Seeds Acuminate. Fairly plump. Fairly straight or slightly curved. Rather light greyish-brown.

Flesh Creamy white. Slightly coarse-textured. Fairly juicy. Firm and crisp. Skin tough.

Aroma Very slight.

Flowers Pollination group 3.

Leaves Medium to small. Oval to broadly oval. Bluntly serrate, or bi-serrate. Medium thick. Flat not undulating. Some slightly upward-folding. Slightly downward-hanging. Not very downy.

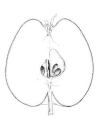

<table>
<tr><td>

MONARCH

Season   November to January
Picking time   Mid September

A fairly well known late culinary apple of English origin that was raised in 1888 by Messrs Seabrook of Chelmsford in Essex and introduced by them in 1918. The parents of this apple were Peasgood Nonsuch × Dumelow's Seedling. It is a very good cooker which cooks to a fine yellow fluff with plenty of juice. It is acidic. The trees are fairly hardy and they are suitable for growing in the West as they show a good resistance to scab. The trees are vigorous, upright-spreading and spur-bearers. The cropping is regular and heavy but with a biennial tendency. It is listed by several specialist nurseries.

Size Large, 80 × 70mm (3⅛ × 2¾″) to 73 × 58mm (2⅞ × 2¼″).

Shape Round-conical to flat-round. Broad flattened base, flattened usually narrower apex. Fairly symmetrical. No ribs, or slight trace of broad ribs. Slightly five-crowned at apex. Sometimes a hair line present at apex. Regular.

Skin Pale greenish-yellow (1C) becoming pale yellow (8C). Slightly to half flushed with greyish-red (179A) to pinkish-red (47A). Flush can be sparse and mottled or densely mottled if exposed to the sun. Some indistinct stripes either of the flush colour or deeper crimson (53B). Lenticels conspicuous green or ochre russet dots on yellow skin and grey russet dots surrounded by red or grey on flush. Skin smooth becoming greasy if stored.

Stalk Medium to fairly stout (3–3.5mm). Short to fairly short (10–15mm). Often fleshy. Usually level with base or protruding slightly beyond.

Cavity Deep or medium. Rather narrow. Can be lipped. Partly lined with fine cinnamon-brown russet which can streak and scatter out over base.

Eye Quite large. Partly or fully open. Sepals very broad, often separated at base, somewhat flat convergent or erect convergent with tips usually reflexed. Not very downy.

Basin Medium width and depth. Slightly ribbed.

Tube Quite broad funnel-shaped.

Stamens Basal.

Core line Basal, clasping.

Core Median. Axile, closed or open.

Cells Ovate.

Seeds Acute. Fairly plump. Straight or curved.

Flesh White. Coarse-textured. Juicy. Tender.

Aroma Almost nil, slightly sharp.

Flowers Pollination group 4. Biennial.

Leaves Large. Broadly oval. Crenate. Medium to thin. Slightly undulating. Very slightly upward-folding. Dark green. Slightly downy.

</td><td>

JONATHAN

Season   November to January
Picking time   Early October

This late dessert apple was first recorded in an article written by Judge J. Buel to the New York Horticultural Society in 1826. The apple originated on a farm at Woodstock, Ulster County, New York, where it was raised by Philip Rick, and thought to be a seedling of Esopus Spitzenberg open pollinated. Jonathan Hasbrouck brought the apple to the attention of Judge Buel and it was named after him. It is grown commercially in many countries and in the UK. The trees are very susceptible to mildew but show some resistance to scab. They make weak, weeping trees which produce spurs freely. The cropping is heavy and regular. The fruits are sweet with weak but pleasant flavour somewhat like pear-drops. It is listed by a few specialist nurseries.

Size Small to medium, 67 × 61mm (2⅝ × 2⅜″).

Shape Round to oblong, or rather conical. Flattened at base and apex. Ribs fairly distinct, angular or well rounded. Slightly five-crowned at apex. Can be flat-sided. Slightly irregular. Symmetrical.

Skin Pale whitish-green (145A) becoming pale greenish-yellow (3C). Quarter to three quarters flushed with intense crimson-red (46A) or thinner paler red (45C). A few indistinct, short, broken stripes of a deeper crimson (53A). Surface slightly hammered. Lenticels very conspicuous tiny dark grey-brown dots on flush or green dots on yellow skin. There can be a hair line present. Skin smooth, dry and quite shiny. A few small grey-brown or greyed-gold russet patches.

Stalk Slender (2 mm). Short to medium length (8–18mm). Within cavity or protruding beyond.

Cavity Deep and fairly narrow. Usually partly or completely lined with fine greyed-gold russet which can streak out a little over base.

Eye Small. Closed. Sepals broad based, short, erect convergent, sometimes pinched in. Fairly downy.

Basin Fairly narrow and deep. Usually russet free. Slightly ribbed sometimes slightly puckered.

Tube Funnel-shaped or deep cone.

Stamens Median.

Core line Basal, clasping or median.

Core Median. Axile, slit.

Cells Obovate.

Seeds Numerous. Fairly large. Acute, longish oval and bluntly pointed. Straight or curved.

Flesh White tinged slightly green. Tender. Fine-textured. Skin tough and chewy. Fairly juicy.

Aroma Almost nil.

Flowers Pollination group 3.

Leaves Quite small. Acute. Serrate often deeply cut. Thin. Some slightly undulating otherwise flat. Upward-folding. Light greyish-green. Very downy.

</td></tr>
</table>

SPARTAN

Season   October to February
Picking time   Early October

This McIntosh type dessert apple is of Canadian origin. It was raised in 1926 at the Dominion Experiment Station, Summerland, British Colombia, by R.C. Palmer from McIntosh × Yellow Newtown Pippin. The original tree first fruited in 1932 and the cultivar was introduced in 1936. It is fairly important commercially. The fruits are small unless thinned and the trees fed generously. The flavour is fairly rich, sweet and juicy, but I found it rather dry by January. The trees are of moderate vigour, upright-spreading in habit and are spur-bearers, producing spurs freely. They are rather prone to canker but are suitable for growing in the North and West. It is widely listed by fruit tree nurserymen.

Size Medium, 64 × 58mm (2½ × 2¼″).
Shape Round-conical. Frequently lop-sided. Rather indistinct ribs, slightly more distinct towards apex. Often five-crowned at apex. Fairly regular.
Skin Pale green (144C) becoming pale yellow (7D). Three quarters to almost completely flushed crimson (53A), or dense purplish-crimson (187B) with some indistinct thin broken stripes of purple (187A) or brownish crimson (185A).Considerably bloomed, some with streaky scarf skin at base. Lenticels distinct tiny whitish or ochre russet dots on flush, and indistinct tiny greenish-grey or greyed-pink dots on yellow or green skin. Skin very smooth and dry.
Stalk Variable. Fairly slender (2.5mm) to medium or stout (3–4mm). Medium length (15–20mm).
Cavity Deep and rather narrow. Often lipped. Can be some fine greenish-ochre russet.
Eye Medium to small. Closed or partly open. Sepals erect, sometimes convergent. Sepals sometimes separated at base with tips reflexed. Very downy.
Basin Medium depth and width. Can be pinched looking. Ribbed and puckered, sometimes beaded.
Tube Deep slightly funnel, almost cone-shaped.
Stamens Median.
Core line Almost basal, clasping.
Core Median. Axile or abaxile.
Cells Obovate.
Seeds Acute or acuminate.
Flesh White. Firm but tender. Fine-textured. Skin slightly tough.
Aroma Strong. Sweetly scented.
Flowers Pollination group 3.
Leaves Medium size. Oval. Serrate. Medium thick. Flat not undulating. Mid grey-green. Very downy.

BESS POOL

Season   November to February
Picking time   Early October

This old dessert apple was found in a Nottinghamshire wood by Bess Pool, the daughter of the village innkeeper. The tree became known and grafts were taken. It was introduced by Mr J.R. Pearson, a nurseryman of Chilwell, whose grandfather procured the grafts. The first record of this apple was in 1824.

The fruit is rich in flavour with a nice balance of sugar and acid, though rather dry. Trees are shy bearers at first but become more productive with age. The trees are moderately vigorous, upright-spreading and do not produce spurs very freely, being slightly inclined to tip-bearing. It is a useful variety to grow in areas prone to late frosts due to the very late flowering. Trees have a very limited availability.

Size Medium 67 × 58mm (2⅝ × 2¼″).
Shape Round-conical. Distinctly flattened at base and apex. Indistinct broad ribs. Symmetrical or slightly lop-sided. Fairly regular.
Skin Greenish-yellow (150B). Quarter to three quarters flushed with mottled and dotted brownish-red (178A) to denser greyed-red (180A). indistinct, broken stripes of purplish-brown (183B). The skin is partly covered with a layer of thin scarf skin giving it a rather milky appearance. Variable amounts of slightly scaly grey-brown russet. Lenticels fairly conspicuous whitish or grey dots with an areola of greyed-red on the yellow skin and purplish-red on the flush. Skin dry, becoming greasy.
Stalk Stout (4mm), short (10–15mm). Level with base. Some have a fleshy knob at the spur end.
Cavity Medium depth and width. Regular. Partly or completely lined with grey or grey-brown russet, sometimes scaly, which can streak over base.
Eye Medium size. Partly open. Sepals erect convergent with tips reflexed. Stamens often visible. Very downy.
Basin Medium depth. Medium width. Five prominent beads. There can be some grey-brown russet.
Tube Funnel-shaped, or almost cone-shaped.
Stamens Median.
Core line Median towards basal.
Core Median. Axile, open or abaxile.
Cells Ovate.
Seeds Acute to obtuse. Plump or fairly plump.
Flesh White sometimes tinged red. Firm but tender. Slightly coarse-textured. Rather dry.
Aroma Slight, sweetly scented.
Flowers Pollination group 6.
Leaves Fairly large. Acute to narrow acute. Very small serrations. Medium thick. Flat or slightly undulating. Slightly upward-folding. Mid green. Slightly downward-hanging. Not very downy.

ROSEMARY RUSSET

Season   November to March
Picking time   Late September to early October

This is a high quality dessert apple. The origin is unknown. It was first described in 1831 and was listed in the catalogue of the 1888 Apple and Pear Conference. It is one of the nicest russets and has a brisk, sweet yet sour flavour with plenty of juice. It is especially good in a warm season. It is rather too small for commercial appeal. The trees are moderately vigorous, upright-spreading and produce spurs freely. The cropping is moderate. It is fairly widely listed by nurserymen.

Size Medium, 64 × 54mm (2½ × 2⅛″).
Shape Conical. Broad and flattened at base tapering to a narrow flattened apex. Slightly five-crowned at apex. Sometimes lop-sided. Fairly distinct ribs. Can be a little flat-sided. Rather irregular.
Skin Pale greenish-yellow (150B). Some fruits unflushed, others slightly to half flushed with orange-ochre (163A), greyed-orange (168A) or greyed-brown (172A). Some indistinct short greyed-red stripes (180A). Partly covered with sparse or dusted fine greenish-ochre russet with a larger area of pale cinnamon-brown russet at apex. Lenticels conspicuous quite large grey-brown russet dots. Skin smooth and dry.
Stalk Medium thick to quite stout (3–3.5mm), occasionally fairly slender (2.5mm). Long (24–28mm). Protrudes well beyond base.
Cavity Fairly wide and deep. Sometimes lipped. Skin usually remains green. Lined with fine grey russet which streaks out over base.
Eye Small to medium. Closed or slightly open. Sepals erect convergent with tips slightly or well reflexed. Can be squashed looking. Fairly downy.
Basin Medium to rather narrow. Medium depth. Slightly ribbed and puckered. Partly or completely russetted with fine pale grey-brown russet.
Tube Conical or funnel-shaped.
Stamens Marginal or median.
Core line Basal clasping or almost basal.
Core Median. Axile.
Cells Variable, ovate or obovate.
Seeds Acute. Large and numerous. Wide, rather flattish, bluntly pointed.
Flesh Pale creamy-white. Firm but not hard. Fine-textured. Juicy.
Aroma Very slight rather acid.
Flowers Pollination group 3.
Leaves Long narrow acute. Medium size. Serrate. Medium thick. Flat not undulating. Slightly upward-folding. Darkish green. Slightly downy.

BARNACK BEAUTY

Season   December to March
Picking time   Late September

This is a very late dessert apple raised in about 1840 and it was reported in the *Gardener's Chronicle* of 1899, that the parent tree, was still growing in a garden in the village of Barnack, four miles from Stamford, in Northants. The variety was introduced in about 1870 by W. & J. Brown of Stamford near Peterborough. It received the Award of Merit from the RHS in 1899, a First Class Certificate in 1909. The trees are very hardy and will grow right up into the North and they show some resistance to scab. They are moderately vigorous, spreading in habit and tip-bearers. The fruit is sub-acid, juicy and very crunchy with a nice refreshing flavour and a good balance of sugar and acid. The cropping is moderate. Listed by one or two specialist nurseries.

Size Medium, 67 × 58mm (2⅝ × 2¼″).
Shape Round to oval. Well rounded at base, flattened at apex. Symmetrical or a little lop-sided. Slight trace of well rounded ribs. Fairly regular.
Skin Yellowish-green (145A) to pale greenish-yellow (3B) becoming yellow (6B). Up to half speckled and dotted with scarlet (42A), the overall appearance is rather orange-red (34A). Fairly distinct broken stripes of crimson-red (46A). Some slight patches and specks of grey-brown russet. Lenticels conspicuous as slightly raised grey-brown russet dots. Surface dry and bumpy. There can be a hair line present.
Stalk Medium thick (3mm). Long (20–25mm).
Cavity Narrow and fairly shallow. Lined with slightly scaly ochre-brown russet that spreads out over base. Can be lipped.
Eye Large, partly to fully open. Sepals broad based, convergent or erect, sometimes separated at base, with tips reflexed. Stamens usually present. Slightly downy.
Basin Wide and fairly shallow. Regular. Slightly puckered. Some golden-brown, sometimes scaly grey-brown, russet usually around apex spreading into basin.
Tube Funnel-shaped, sometimes broad or deep.
Stamens Marginal.
Core line Median joining at funnel neck.
Core Median. Axile, usually closed, can be open.
Cells Roundish ovate. Very slightly tufted.
Seeds Acute or acuminate. Fairly plump. Straight.
Flesh Yellowish. Fine-textured. Juicy and hard. Skin tough.
Aroma Almost nil.
Flowers Pollination group 4.
Leaves Medium to smallish. Acute. Serrate. Rather thin. Flat not undulating. Mid green. Slightly downy.

HOWGATE WONDER

Season   October to March
Picking time   Early October

A very large late culinary apple which was raised in 1915–16 by Mr G. Wratton of Howgate Lane, Bembridge, Isle of Wight. It was introduced in 1932 by Stuart Low & Co. of Enfield in Middlesex and in 1949 it received the Award of Merit from the RHS. It is a heavy and regular cropper when fully established producing an attractive fruit that stores well. It is a useful pollinator for Bramley's Seedling. The trees are vigorous and produce spurs freely and being hardy are suitable for growing in the North. The fruit is only fair in quality, sub-acid with not much flavour. The flesh becomes a rather dirty yellow when cooked and breaks up almost completely. Trees are widely available from nurserymen.

Size Very large, 86 × 72mm (3⅜ × 2⅞″).
Shape Round-conical. Flattened at base tapering to a flattened apex. Fairly distinct broad ribs which can be angular and terminate in five pronounced crowns at apex. Fruit can be a little flat-sided. Fairly regular. Symmetrical or lop-sided.
Skin Light yellowish-green (144B) to greenish-yellow (151A). Slightly to half flushed with thin orange-brown (175C) with short broken stripes of the same colour or a darker greyed-red (178B). Usually some patchy scarf skin at base sometimes slightly on cheeks, which can make the flush appear more greyed-purple (184A). Usually no russet on cheeks. Lenticels indistinct white or grey dots surrounded by a circle of scarf skin which become more numerous at apex and enter basin. Skin very smooth and dry.
Stalk Very short to short (8–13mm) and stout (5mm). Often fleshy. Within cavity.
Cavity Wide and fairly deep. Frequently lipped with the stalk enveloped in the lip and sunk deep into cavity. Usually green with conspicuous dots of scarf skin and partly or completely lined with light brown russet which can streak over base.
Eye Large. Closed or half open. Sepals convergent, broad at base with tips reflexed. Very downy.
Basin Wide and deep. Ribbed and slightly puckered. Irregular.
Tube Funnel-shaped.
Stamens Median.
Core line Median. Sometimes two.
Core Median. Axile.
Cells Obovate. Slightly tufted.
Seeds Acute. Broad and fairly plump. Straight.
Flesh Creamy-white. Firm. Fine-textured.
Aroma Very slight when cut, otherwise nil.
Flowers Pollination group 4.
Leaves Medium. Broadly acute. Serrate. Medium thick. Slightly undulating. Dark green. Downy.

IDARED

Season   November to April
Picking time   End of October to early November

This American dual-purpose apple is grown commercially in the UK and often used as a pollinator. It was raised by Leif Verner at the Idaho Agricultural Experiment Station at Moscow, Idaho from Jonathan × Wagener. It was selected in 1935 and introduced in 1942. It is fairly important commercially because of its late keeping properties. The trees are fairly hardy, moderately vigorous, upright-spreading and produce spurs very freely. The cropping is good. The fruit is rather watery with weak but quite pleasant flavour. It is widely listed by nurserymen. Idared shows some resistance to scab.

Size Medium large, 70 × 58mm (2¾ × 2¼″).
Shape Flat-round. Flattened at base and apex. Usually fairly symmetrical. Fairly distinct ribs. Can be a little flat-sided. Fairly regular.
Skin Pale greenish-yellow (150C) becoming whitish-yellow (5D). Quarter to three quarters flushed bright crimson-red (46A) to paler pinky-red (45C) on less coloured fruits. Rather indistinct, short, broken stripes of crimson-red (46A) to purplish-crimson (187B) on well coloured fruits. Usually russet free on cheeks but can have some russetted injury patches. Lenticels rather inconspicuous tiny whitish russet dots. Skin smooth, very shiny and dry.
Stalk Fairly slender (2.5mm). Usually fairly long (18–27mm). Protrudes well beyond base.
Cavity Narrow and deep, often lipped. Usually partly or wholly lined with golden-brown russet overlaid with some fine grey scaly russet, which can streak out a little over base.
Eye Fairly small. Closed or slightly open. Sepals small, erect convergent or connivent with tips reflexed or broken off. Moderately downy.
Basin Medium depth. Medium to narrow. Ribbed and slightly puckered. Usually a dusting or scattered patches of brown or grey-brown russet which can spread over apex.
Tube Deep slightly funnel-shape or narrow cone.
Stamens Marginal or slightly towards median.
Core line Median.
Core Median. Axile.
Cells Round to roundish obovate.
Seeds Acuminate to acute. Moderately plump.
Flesh Very white or white tinged green, sometimes tinged pink. Firm. Fairly fine-textured. Juicy and crisp. Skin rather tough.
Aroma Nil.
Flowers Pollination group 2.
Leaves Medium size. Acute. Serrate. Medium thick. Slightly upward-folding, or slightly downward-folding. Mid to slightly blue-green. Quite downy.

BELLE DE PONTOISE

Season   November to March
Picking time   Mid October

As its name suggests, this apple originated in France, raised by Rémy père of Pontoise in 1869 from a seed of Emperor Alexander. It is not grown commercially in the United Kingdom and is primarily a garden and exhibition fruit. The trees are moderately vigorous, upright-spreading and spur-bearers, producing spurs freely. The cropping is good though it tends to be biennial and the fruit is sub-acid with a rather weak though pleasant flavour. The flesh becomes rather brownish-yellow when cooked and breaks up completely though not to a fluff. Not listed by nurseries.

Size Very large, 86 × 64mm (3⅜ × 2½″).

Shape Flat-round, sometimes slightly conical. Very broad and flattened at base, flattened at apex. Symmetrical. Rather irregular with fairly prominent ribs which are sometimes angular. Can be flat-sided or even slightly concave. Occasionally five-crowned at apex.

Skin Pale green (144C) becoming pale greenish-yellow (5C). Quarter to three quarters flushed with rather dull reddish-brown (178B) to fairly bright carmine (53A). Fairly broad broken stripes of deep brownish-crimson (185A) rather indistinct on the darker flush but sometimes appearing over the green or yellow skin. Some streaky scarf skin at base. Lenticels fairly conspicuous pale grey russet dots on flush, grey-brown russet dots on green or yellow skin. Skin very smooth and shiny becoming greasy.

Stalk Fairly stout to stout (3.5–4.5mm). Medium length (15–20mm). Protrudes beyond base.

Cavity Deep and very wide. Can be lipped. Lined with greenish-ochre and some slightly scaly pale brown russet which may spread out over base.

Eye Large. Partly to half open. Sepals broad based, erect convergent or flat convergent with tips fairly erect or broken off.

Basin Deep and fairly wide to wide. Slightly ribbed, sometimes slightly puckered. Very downy.

Tube Wide and fairly shallow. Cone-shaped.

Stamens Median or towards basal.

Core line Basal, clasping, sometimes meeting.

Core Median. Axile.

Cells Round or slightly obovate.

Seeds Quite large. Acute. Wide and fairly plump.

Flesh White. Firm and fairly juicy. Slightly coarse-textured.

Aroma Slight and sweet.

Flowers Pollination group 3. Biennial.

Leaves Fairly large. Broadly acute. Serrate or broadly serrate. Medium thick and rather leathery. Undulating. Dark rather yellowish-green. Slightly downward-hanging. Undersides slightly downy.

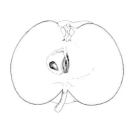

NEWTON WONDER

Season   November to March
Picking time   Mid October

This is a very large, very colourful, late to very late culinary apple. It was raised in England by Mr Taylor of King's Newton, Melbourne, Derbyshire, said to be from Dumelow's Seedling × Blenheim Orange. It was introduced by Messrs Pearson & Co. in about 1887 and in that year received a First Class Certificate from the RHS. The tree is too vigorous for the small garden and should be grafted on to a dwarfing rootstock if required for limited areas or restricted forms. It is fairly hardy and suitable for growing in the North and West, but overlarge fruits are prone to bitter pit. It is grown on a small scale commercially and is widely available from nurseries. The fruits are sub-acid with a good full flavour and cook to a yellow fluff.

Size Very large, 89–92 × 70mm (3½–3⅝ × 2¾″).

Shape Flat-round. Symmetrical or lop-sided. Well flattened at base and apex. Sometimes a slight trace of ribs. Fairly regular.

Skin Pale yellow-green (145B) to greenish-yellow (2B) becoming fairly bright yellow (7B) to golden yellow (13B) on ripening. Quarter to three quarters flushed with brownish-red (173A), to very bright red (46B) on ripening, with some golden yellow skin showing through giving it a slightly orange-red appearance from a distance. Fairly distinct broad broken stripes of crimson-red (46A). Lenticels conspicuous grey-brown or greenish-grey russet dots. Skin smooth or slightly bumpy, becoming greasy.

Stalk Stout to very stout (4–6mm). Short (10–15mm). King fruits fleshy where the stalk joins the fruit. Level with base or protrudes slightly.

Cavity Fairly shallow and fairly narrow. Can be lipped. Russet free or a small amount of very fine greenish-ochre russet within cavity.

Eye Large. Fully open. Sepals flat convergent with tips reflexed or broken off. Slightly downy.

Basin Fairly wide and deep. Regular. Slightly puckered. There can be some fine grey-brown russet.

Tube Deep funnel-shaped.

Stamens Median or towards marginal.

Core line Median.

Core Median. Axile.

Cells Fairly small. Round or slightly ovate.

Seeds Small. Acute. Plump. Straight or curved.

Flesh Creamy, slightly yellowish-white. Firm and crisp. Slightly coarse-textured. Fairly juicy.

Aroma Sweet but slight.

Flowers Pollination group 5. Biennial.

Leaves Fairly large. Broadly oval or almost round. Serrate or bi-serrate. Medium thick. Flat. Slightly upward-folding. Mid grey-green. Fairly downy.

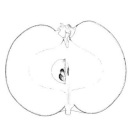

DUMELOW'S SEEDLING

Season   November to March
Picking time   Mid October

This very late culinary apple was raised by Mr Dumeller (pronounced Dumelow), a farmer at Shakerstone, Leicestershire thought to be from seed of the Northern Greening. The original tree was known to be growing in 1800. It was known in that area as Dumelow's Crab but was introduced in 1819 or 1820 by the Turnham Green Nursery as Wellington Apple. This apple used to be widely grown for the mincemeat trade because of its good flavour. The trees are moderately vigorous, fairly spreading and produce spurs freely. They are fairly hardy and suitable for growing in the North. Lenticels are very noticeable on the wood particularly on the young growth. The cropping is good and the fruits very acid. They cook to a pale cream puree. Trees have a very limited availability from nurserymen.

Size Large, 77 × 61mm (3 × 2⅜″).
Shape Flat-round. Flattened at base and apex. Symmetrical. No ribs or slight trace. Regular.
Skin Pale yellowish-green (between 144D and 150C), becoming pale yellow (6C). Some fruits quarter to half delicately blushed with pinkish-orange (26C) with a few short pinkish-red stripes (42B). Lenticels numerous, white or pinkish-brown russet dots in a circle of green when not on flush. There can be some specks or small patches, of fine golden russet. Some scarf skin at base. There can be hair line. Surface smooth or slightly textured and somewhat pitted. Skin greasy.
Stalk Fairly stout (3.5mm). Short (11–14mm). Level with base or protruding slightly beyond.
Cavity Narrow. Medium depth. Partly lined with fine greenish-gold or grey-brown russet which can scatter over base. Regular. Occasionally lipped.
Eye Large, fully open. Sepals fairly long, erect, separated at base with tips reflexed or broken off. Slightly downy.
Basin Medium to fairly wide and rather shallow. Sometimes a hint of ribs and sometimes a little wrinkled. There can be some golden-brown russet.
Tube Broad funnel-shaped or cone-shaped.
Stamens Basal.
Core line Basal, clasping.
Core Median. Axile or abaxile.
Cells Roundish obovate.
Seeds Acute. Roundish, plump and often tufted.
Flesh Whitish. Firm crisp and very juicy. Rather coarse-textured.
Aroma Slight.
Flowers Pollination group 4.
Leaves Medium. Acute. Crenate or bluntly serrate. Medium thick. Flat or slightly undulating. Slightly upward-folding. Dark grey-green. Fairly downy.

CORNISH GILLIFLOWER

Season   November to March
Picking time   Mid October

According to Robert Hogg in *The Fruit Manual*, the word Gilliflower is derived from the old French word Girofle, signifying a clove and this apple is said to emit a clove-like fragrance when cut. The original tree was found in about 1800 growing in a cottage garden in Truro, Cornwall. It was brought to the Horticultural Society of London in 1813 by Sir Christopher Hawkins and was awarded the Society's Silver Medal. It has always been considered a high quality dessert apple, possessing a sweet rich flavour. The trees are moderately vigorous, very spreading and tip-bearers making them unsuitable for growing in restricted form. The cropping is light to moderate. It is listed by a few nurserymen.

Size Medium large, 70 × 70mm (2¾ × 2¾″).
Shape Oblong to oblong-conical. Very distinct narrow angular ribs. Sometimes flat-sided. There can be smaller ribs in between the five main ones. Well rounded at base. Distinctly five-crowned at apex. Can be a little lop-sided. Irregular.
Skin Deep yellow-green (144A) to liverish-green (152A) becoming yellow (9A). Slightly to half flushed with rather speckled red (45A) with the yellow (17A) skin showing through. Numerous distinct broad broken stripes of red (46A) to purplish-crimson (187B). Some grey-brown slightly netted, or dusted russet, gives the fruit a rather scruffy appearance. Lenticels very distinct whitish or greenish-grey dots. Skin slightly rough or bumpy and dry.
Stalk Fairly slender to slender (2.5–2mm). Medium length (18mm). Sometimes fleshy. Beyond base.
Cavity Narrow and rather shallow to medium depth. Irregular. Frequently dark rather dirty green and partly lined with grey russet.
Eye Medium size. Closed or slightly open. Sepals erect convergent or connivent with tips reflexed. Rather pinched looking. Fairly downy.
Basin Narrow and fairly shallow. Distinctly ribbed, sometimes puckered. Some fine cinnamon-brown russet, particularly at apex.
Tube Cone-shaped or funnel-shaped.
Stamens Median or marginal.
Core line Median to basal, sometimes two.
Core Median to slightly distant. Axile or abaxile.
Cells Roundish obovate. Tufted.
Seeds Acute. Plump. Dark brown.
Flesh Yellow tinged green towards core. Firm. Fine.
Aroma Fairly distinct and sweet.
Flowers Pollination group 4.
Leaves Rather small. Narrow acute. Bluntly serrate. Thin. Upward-folding. Mid bluish-green. Downy.

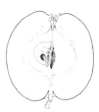

HAMBLEDON DEUX ANS

Season   November to April
Picking time   Late September

This old dual-purpose apple is of English origin having come from Hambledon in Hampshire in about 1750. The fruit has a fair flavour, sweet but with a nice balance of sugar and acid. It cooks yellow and breaks up almost completely but without much juice. The trees are very vigorous, upright-spreading and partial tip-bearers and they are rather prone to bitter pit. The cropping is erratic. It does not appear to be listed by nurserymen today.

Size Large, 77 × 64mm (3 × 2½").
Shape Round-conical to slightly oblong. Flattened at base and apex. Can be regular but often irregular with some well pronounced ribs. Sometimes flat-sided. Often lop-sided.
Skin Deep grass-green (144A) becoming greenish-yellow (151B, a little too yellow). Slightly to half flushed with reddish-brown (172A) to brownish-ochre (163A). Broad broken stripes of dull brownish-purple (176A). A considerable amount of dense sometimes rather striped scarf skin at base frequently extending up the sides of the fruit appearing as a rather mottled layer of greenish-white over the green skin and mottled purplish-white over the flush. Skin also mottled with varying amounts of rather dirty grey-brown russet making the skin very slightly rough to the touch, otherwise smooth. Lenticels inconspicuous pinkish or grey-brown russet dots. Skin dry.
Stalk Medium to stout (3–4mm). Short (10–15mm). Within cavity or slightly beyond.
Cavity Medium depth, fairly wide. Often lipped. Usually dark green and lined with grey-brown slightly scaly russet which often spreads over base.
Eye Smallish. Closed or slightly open. Sepals broad, connivent sometimes rather pressed together with tips reflexed. Very downy.
Basin Medium width and depth. Irregular. Ribbed and a little beaded. Sometimes a little grey russet.
Tube Cone-shaped.
Stamens Median.
Core line Basal, clasping.
Core Median. Abaxile, wide open.
Cells Obovate. Tufted.
Seeds Acute. Rather starved looking. Longish and blunt.
Flesh Creamy-white. Firm. Crisp. Slightly coarse-textured. Dry.
Aroma Very slight, rather scented.
Flowers Pollination group 3.
Leaves Large. Acute. Serrate. Medium thick. Upward-folding. Slightly undulating. Dark grey-green. Undersides very downy.

ANNIE ELIZABETH

Season   December to June
Picking time   Mid October

This fairly old English culinary apple was raised by Samuel Greatorex at Knighton in Leicester about 1857. It received a First Class Certificate from the RHS in 1866 and was introduced around that time by Messrs Harrison & Son of Leicester. It is named after the two daughters of Mr Thomas Harrison, proprietor of the nursery. It used to be grown commercially and it is still quite widely listed by nurserymen. The trees are fairly hardy and can be grown in all areas. They are moderately vigorous, upright in habit and spur-bearers. The cropping is uncertain. The fruit is fairly acid with a good flavour and cooks to a pale greenish-yellow fluff.

Size Large, 79 × 63mm (3⅛ × 2½").
Shape Roundish-oblong. Slightly flattened at base and apex. Fairly well defined sometimes slightly angular ribs. Rather irregular. Can be slightly five-crowned at apex. Frequently lop-sided.
Skin Light yellowish-green (145A) to yellow (2B). Some fruits quarter to half flecked, speckled and striped with greyed-red (179A), others more densely flushed greyed-red (179A) to pinkish-red (47A). Fairly numerous short broad stripes of greyed crimson (53B). Lenticels fairly conspicuous pale grey-brown, green or ochre dots. Some fine scarf skin at base. There can be a hair line. Surface hammered. Skin smooth and fairly greasy.
Stalk Medium thick (3mm). Short (9–18mm), can be fleshy. Level with base or just beyond.
Cavity Medium depth. Fairly wide. Stalk set well down within. Dark green. Partly or completely lined with fine pale brown or ochre russet. Lenticels enter cavity as large greenish-white dots.
Eye Quite large. Closed or partly to half open. Sepals broad, quite long and tapering, erect or convergent, sometimes connivent and rather squashed-in looking. Fairly downy.
Basin Medium width to rather narrow. Deep. Ribbed and slightly puckered. Some greenish-brown or ochre-brown russet.
Tube Deep, cone-shaped, can enter core cavity.
Stamens Median or basal.
Core line Median or towards basal.
Core Distant. Abaxile.
Cells Obovate, sometimes roundish.
Seeds Acute. Plump. Straight or curved.
Flesh Creamy-white. Crisp but fairly tender. Rather coarse-textured. Fairly juicy.
Aroma Fairly strong and sweet.
Flowers Pollination group 4.
Leaves Large. Broadly acute. Deeply serrate, or bi-serrate. Medium thick. Slightly undulating and upward-folding. Dark blue-green. Fairly downy.

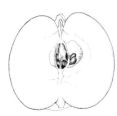

BELLE DE BOSKOOP

Season   December to April
Picking time   Early October

This is a fairly old dual-purpose apple which came from Holland. It was found at Boskoop, near Gouda by K.J.W. Ottolander in 1856. It was introduced to England and received the Award of Merit from the RHS in 1897. It is grown commercially in Holland and Germany but not in the United Kingdom. The parentage is not certain but it is thought to be a bud sport of Reinette de Montford. It is listed by a few fruit tree nurserymen today. The trees are vigorous, upright-spreading, spur-bearers and triploid. The cropping is good to heavy. It is suitable for growing in the West. The fruit is slightly sweet and quite acid with a fairly rich flavour but rather dry. It cooks to a golden yellow fluff with a fair, sub-acid flavour and rather woolly.

Size Medium large, 73 × 67mm (2⁷⁄₈ × 2⁵⁄₈″).
Shape Round-conical. Often a little lop-sided. Slight trace of ribs, with one more prominent. Can be very irregular. Flattened at base and apex.
Skin Light greenish-yellow (4A) to light yellow (7B). Slightly to quarter mottled, dotted and striped with bright red (45A) which with the orange skin showing through appears rather orange-red from a distance. About half to three quarters covered with fine ochre russet, sometimes netted, with base completely covered. Lenticels fairly conspicuous slightly raised grey or greenish russet dots. Skin dry and hammered.
Stalk Stout (4mm). Medium length (17–24mm). Protrudes slightly beyond base.
Cavity Medium width and quite deep with stalk well sunk within. Completely lined with fine ochre russet and overlaid with some brown scaly russet, which extends over shoulder and entire base.
Eye Large. Partly to half open. Sepals green, broad based long and convergent with tips up to half reflexed. Not very downy.
Basin Medium width to fairly wide. Deep. Ribbed and irregular. Partly lined with fine ochre russet.
Tube Rather wide funnel-shape.
Stamens Median.
Core line Median sometimes towards basal.
Core Median. Abaxile.
Cells Round or elliptical. Or slightly obovate.
Seeds Acute. Long and bluntly pointed. Rather sparse. Some plump and some rather starved.
Flesh Cream tinged slightly green. Firm. Coarse-textured. Rather dry.
Aroma Nil.
Flowers Pollination group 3. Triploid.
Leaves Medium to large. Broadly oval. Finely and sharply serrate. Medium thick. Flat not undulating. Mid greyish-green. Undersides fairly downy.

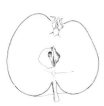

WILLIAM CRUMP

Season   December to February
Picking time   Mid October

A high quality and attractive dessert apple which was raised by W. Crump of Madresfield Court Gardens, Malvern, Worcestershire. Mr Crump exhibited this apple on 22 December 1908 and was awarded the RHS Award of Merit. In January 1910 the fruit was exhibited by Earl Beauchamp of Madresfield, and was awarded a First Class Certificate. It was introduced by Messrs Rowe of Worcester. It is not grown commercially in the United Kingdom today. The trees are vigorous, upright in habit and spur-bearers. The cropping is light to moderate. The fruit has a rich flavour with a nice balance of sugar and acid and is crisp and juicy. Trees are available from one or two specialist nurseries.

Size Medium large, 70 × 58mm (2¾ × 2¼″).
Shape Round-conical. Distinctly flattened at base and apex. Slightly five-crowned at apex. Can be a little waisted near apex. Rather indistinct ribs sometimes with one very broad more prominent one. Usually fairly regular. Fairly symmetrical.
Skin Yellowish-green (151D) becoming yellow (7B). Half to almost completely covered with brownish-crimson flush (185A) which is crimson-red (46A) at the extremities. Rather indistinct broken stripes of brownish-crimson (185A). Scattered with dots, patches and little scratches of grey russet. There can be some scarf skin at base. Lenticels fairly conspicuous grey-brown russet dots. Skin smooth and slightly greasy.
Stalk Medium thick (3mm). Short to medium length (13mm). Set deep into cavity.
Cavity Very deep cone. Fairly narrow, or broadening out and becoming wide at base. Lined with grey-brown scaly russet.
Eye Medium size, closed or partly open. Sepals small, erect convergent with tips reflexed. Very downy. Stamens often visible.
Basin Shallow. Medium width. Trace of ribbing.
Tube Long, narrow funnel shape.
Stamens Almost marginal.
Core line Median.
Core Median. Axile, closed or open, or abaxile.
Cells Ovate.
Seeds Rather large. Acute, broad and fairly plump, bluntly pointed. Straight not curved.
Flesh Cream deepening in colour towards the flush. Firm. Fairly fine-textured. Crisp and juicy.
Aroma Very slight and sweet.
Flowers Pollination group 5.
Leaves Medium to largish. Acute to broadly acute. Serrate. Medium to rather thin. Some slightly undulating otherwise flat. Slightly upward-folding. Mid green. Undersides very downy.

— CLAYGATE PEARMAIN —

Season   December to February
Picking time   Early October

This highly esteemed dessert apple was found growing in a hedge. It was discovered by John Braddick and the hedge was near his home in Claygate, a hamlet in the parish of Thames Ditton, Surrey. It was found before 1822 but in that year, on 19 February, John Braddick sent specimens to the meeting of the Horticultural Society of London for the first time. It was decided that 'it is unquestionably a first rate dessert apple'. In 1901 it received the Award of Merit from the RHS and, in 1921, a First Class Certificate. The fruit has a rich almost nutty flavour, with a good balance of sugar and acid and a very refreshing zest. The trees are moderately vigorous, upright-spreading and partial tip-bearers; the cropping is abundant. This apple is still listed by several specialist nurseries.

Size Medium large, 70 × 67mm (2¾ × 2⅝″).

Shape Oblong-conical. Flattened at base and apex. Symmetrical or slightly lop-sided. Slight trace of well rounded ribs. Fairly regular. Surface bumpy.

Skin Rather dull green (145A) becoming yellow-green (153B). Slightly to half covered with greyed-orange flush (164A) deepening to greyed-red (179A) nearest the sun. Short broken stripes of red (46A). Variably covered with grey, slightly scaly russet (199D) the scaling giving it a silverish nnd even pinkish tinge. Skin dry and slightly rough.

Stalk Medium to stout (3–4mm). Short to medium (10–18mm). Level with base or protruding beyond.

Cavity Medium width. Medium to fairly deep. Sometimes lipped. Lined or partly lined with grey sometimes scaly russet.

Eye Large. Open. Sepals broad and long, erect convergent with tips reflexed. Some stamens visible. Not very downy.

Basin Wide and moderately deep. Slightly ribbed. Partly or completely russetted.

Tube Short funnel-shaped.

Stamens Median.

Core line Basal, clasping.

Core Median. Axile.

Cells Elliptical, sometimes obovate.

Seeds Acuminate. Very long and sharply pointed. Rather thin.

Flesh Whitish, tinged slightly green. Firm, crisp and juicy.

Aroma Nil, very slight when cut.

Flowers Pollination group 4.

Leaves Medium size. Broadly acute. Deeply and sharply serrate. Medium thick. Flat. Slightly upward-folding. Mid green. Very downy.

—— RED DELICIOUS ——

Season   December to March
Picking time   Mid October

Red Delicious is principally an American apple. It is a coloured clone from the very old variety called Delicious, which is not grown commercially because of its rather dull appearance. It is grown to some extent on the Continent and there are some small commercial plantings in the United Kingdom. However, it requires plenty of sun, hence it does better in warmer climates. It is one of the most widely grown apples in the world, and there are literally hundreds of sports. The trees are moderately vigorous, upright-spreading and spur-bearers. The cropping is good. The fruit is sweet with a fairly good flavour and the skin is very tough. Trees have a very limited availability in the UK.

Size Medium large, 70 × 70mm (2¾ × 2¾″).

Shape Oblong to oblong-conical. Flattened base, rounded shoulders. Very pronounced ribs, especially towards apex, with five high crowns at apex. Rather flat-sided towards apex where it can also be rather waisted. Frequently lop-sided. Irregular.

Skin Pale dull greenish-yellow (150C). Usually entirely covered with crimson (36A) to carmine (53A) to greyed red (178B) on shaded side. Fruit is covered with a dense bloom when on the tree which rubs off and the fruit polishes to an intense shine. Lenticels conspicuous numerous brownish-white dots. No russet. Skin very smooth and dry.

Stalk Fairly stout to stout (3.5–4mm), thickening towards attachment to fruit. Medium to fairly long (19–22mm). Protrudes well beyond base. Can be set at an angle.

Cavity Wide. Medium depth. Sometimes lipped. Frequently some fine dirty grey russet within.

Eye Medium size. Slightly open. Sepals narrow and sharply tapered, erect convergent with some tips reflexed.

Basin Medium width and deep. Prominently ribbed, slightly puckered. Often falls away on one side. Skin in basin usually paler brownish-crimson or dull yellow skin colour. No russet. Fairly downy.

Tube Deep funnel-shape.

Stamens Marginal.

Core line Median.

Core Median. Axile, open.

Cells Obovate. Tufted.

Seeds Acute. Plump. Usually straight.

Flesh Creamy-white tinged green. Firm. Fine-textured. Very juicy.

Aroma Slight, sweet and rather scented.

Flowers Pollination group 3.

Leaves Medium to smallish. Acute. Bluntly serrate. Medium to fairly thick. Flat not undulating. Very slightly upward-folding. Dark green. Fairly downy.

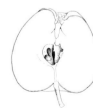

CORNISH AROMATIC	HOLSTEIN

— CORNISH AROMATIC —

Season   December to March
Picking time   Mid October

This apple was first brought to notice by Sir Christopher Hawkins in 1813 and it was thought then to have been known in Cornwall for a great many years. It is not sufficiently fertile for commercial planting but has always been a popular garden variety and is still listed by a few nurserymen. The fruit is sweet with a nice acidic bite, aromatic and slighty spicy, but rather dry: the cropping is good. The trees are vigorous, upright-spreading and make a lot of young growth. They produce spurs freely.

Size Medium, 64 × 58mm (2½ × 2¼″).

Shape Round-conical to oblong-conical. Usually rather lop-sided. Distinctly ribbed particularly towards apex. Distinctly five-crowned at apex. Slightly irregular.

Skin Greenish-yellow (150B) becoming soft creamy yellow (10A). Quarter to half flushed with orange-red (34A). Indistinct short broken stripes of crimson-red (46A). Partly covered with rather dirty grey-brown russet which can be quite liberally dusted or in patches. Lenticels very noticeable and numerous pale greyish-white russet dots or smaller dark-grey russet dots. Skin dry and slightly rough.

Stalk Medium-thick (3mm). Medium length (18–21mm). Extends beyond base.

Cavity Medium width or narrow. Medium to fairly deep. Usually completely lined with fine ochre-green or grey russet which can streak out over base.

Eye Small to medium size. Closed or half open. Sepals erect convergent with tips reflexed. Slightly downy.

Basin Medium depth. Fairly narrow. Distinctly ribbed and some puckering. Sometimes rather pinched-in looking. Usually russetted with patchy pale cinnamon or grey russet. Sometimes beaded.

Tube Cone shaped or slightly funnel-shaped. Usually rather small.

Stamens Median.

Core line Basal, clasping.

Core Median. Axile.

Cells Obovate.

Seeds Acuminate to acute. Fairly plump. Dark brown. Usually straight.

Flesh Yellowish-white or tinged green. Firm, crisp and fine-textured.

Aroma Nil uncut. Sweetly aromatic when cut.

Flowers Pollination group 4.

Leaves Smallish. Acute. Bluntly serrate to crenate. Medium thick. Mostly upward-folding. Very slightly undulating. Mid greyish-green. Undersides fairly downy.

— HOLSTEIN —

Season   November to January
Picking time   Late September

This late dessert apple originated in Germany. It was raised or discovered by a teacher named Vahldik in Eutin, Holstein. It originated in about 1918 and was produced from Cox's Orange Pippin open-pollinated. It is grown on a small scale commercially in the United Kingdom but more so on the Continent, particularly Germany. The fruit is highly aromatic with a Cox-like flavour and sweet with a good balance of acid. It is a delicious apple, one of the best. The trees are very vigorous, wide-spreading in habit and produce spurs freely. It is a triploid. Trees are available from a few specialist nurseries.

Size Medium large, 73 × 67mm (2⅞ × 2⅝″).

Shape Oblong conical to round conical. There can be some slight well rounded ribs and it can be slightly five-crowned at apex. Rounded at base, flattened at apex. Regular. Often lop-sided.

Skin Light yellowish-green (145A–151D) becoming golden yellow (12A). Quarter to three quarters flushed. On ripe fruit the flush appears bright orange from a distance (169C) but is actually tiny dots and stripes of red (45A) over the golden yellow skin. Fairly conspicuous, thin, broken stripes. Small grey russet patches. There may be patchy, rather scruffy scarf skin at base. Lenticels inconspicuous grey russet dots, which are raised making surface bumpy. Slightly greasy.

Stalk Medium to fairly stout (3–3.5mm). Fairly short (9–15mm). Usually protrudes slightly beyond base. Can be set at an angle.

Cavity Medium to narrow and usually lipped. Fairly shallow. Skin green. Completely lined with fine grey-brown russet, which spreads over base.

Eye Large. Wide open. Sepals green, broad, long, erect convergent, sometimes rather flat-convergent, with tips well reflexed. Fairly downy.

Basin Wide and rather shallow. Slightly ribbed. Usually lined with fine or slightly scaly grey-brown russet which can spread over apex.

Tube Funnel-shaped.

Stamens Median, joining at funnel neck.

Core line Median, slightly lower than stamens.

Core Median. Axile, open.

Cells Obovate. Tufted.

Seeds Acute to acuminate. Often rather starved, otherwise plump. Fairly broad and straight.

Flesh Creamy yellow. Slightly coarse-textured. Rather tender but crisp. Juicy.

Aroma Very slight. Stronger when cut.

Flowers Pollination group 3. Triploid.

Leaves Largish. Broadly oval to broadly acute. Serrate. Medium thick. Flat. Margins sometimes curled upwards. Mid blue-green. Very downy.

CRISPIN (MUTSU)

Season   December to February
Picking time   Mid October

Correctly named Mutsu, this apple originated in Japan from Golden Delicious × Indo. It was introduced to the United Kingdom and re-named Crispin in 1968. The apple was raised at the Aomori Apple Experiment Station and first fruited in 1937. It received the Award of Merit from the RHS in 1970. Crispin shows promise commercially as it yields heavily and regularly, the fruits are of good size, and it has the advantage of being a dual-purpose apple. When eaten as a dessert it is fairly sweet with a slight acidity which makes it refreshing. When cooked it stays intact, is pale yellow in colour and has a pleasant though not strong flavour. The trees are very vigorous, spreading and triploid. They produce spurs freely. The cropping is very heavy though with a biennial tendency. Trees are widely available.

Size Large, 77 × 73mm (3 × 2⅞″) or medium large, 73 × 70mm (2⅞ × 2¾″).
Shape Oblong. Definitely ribbed and sometimes flat sided. Ribs can be angular and are more prominent towards apex. Tapers towards apex where it is five-crowned. Symmetrical or slightly lop-sided. Fairly regular.
Skin Yellowish-green (150B) sometimes with a slight flush of greyed-orange (173C–174B). Lenticels conspicuous green, grey-brown or whitish dots, larger towards base. Skin smooth and dry.
Stalk Fairly slender (2.5mm). Long (22–30mm). Extends well beyond base.
Cavity Fairly deep and wide, sometimes lipped. Lined with fine golden-brown russet with silvery sheen, which streaks and scatters over base.
Eye Large. Closed or slightly open. Sepals long and tapering, erect convergent, sometimes slightly separated at base. Fairly downy.
Basin Medium depth and quite wide. Ribbed and slightly puckered.
Tube Cone-shaped.
Stamens Basal.
Core line Quite pronounced. Basal, clasping.
Core Median. Axile, open.
Cells Obovate. Tufted.
Seeds Acute. Slightly curved. Not very plump.
Flesh Creamy-white tinged slightly green. Firm. Fairly juicy. Coarse-textured.
Aroma Nil.
Flowers Pollination group 3. Triploid. Biennal.
Leaves Medium to large. Acute to broadly acute. Sharply and deeply serrate. Medium to rather thin. Flat not undulating. Upward-folding. Mid green. Undersides not very downy.

JOHN STANDISH

Season   December to February
Picking time   Mid to end October

This apple was first brought to notice in 1921 when it was exhibited at the RHS show by Isaac House & Sons of Bristol and received a Highly Commended. It is thought to have been raised by John Standish of Ascot in Berkshire in about 1873. In 1922 it received the Award of Merit from the RHS. It is an attractive scarlet apple the flesh of which is very firm and fairly acid. There is not a great deal of flavour but what there is is quite pleasant. The skin is fairly chewy. The trees are hardy, vigorous, upright in habit and spur-bearers: the cropping is good. It is listed by one or two specialist nurseries.

Size Medium, 64 × 54mm (2½ × 2⅛″).
Shape Round to round-conical. Rounded at base, slightly flattened at apex. Sometimes a trace of one well-rounded rib, otherwise no ribs. Regular. Symmetrical or slightly lop-sided.
Skin Very pale greenish-yellow (2C) becoming very pale whitish-yellow (8B–8C). Half to three quarters flushed with very bright scarlet red (42A), paler on shaded side (42B). The flush can be a deeper blood red (45A) on well coloured fruit. Indistinct, broken stripes of the deeper red (45A) which are not very apparent on the well coloured fruit. Lenticels conspicuous pale grey or greenish-grey dots on flush or grey-brown russet dots on yellow skin. Small amount of scarf skin at base around cavity. Some small grey-brown russet patches.
Stalk Fairly slender to medium (2.5–3mm). Medium to fairly long (15–23mm). Extends beyond base.
Cavity Shallow to medium depth. Medium width. Partly lined or just a dusting of grey-brown russet that can spread on to shoulder.
Eye Fairly small. Open. Sepals short, erect, convergent with tips reflexed. Very downy.
Basin Usually very shallow. Sometimes the eye is almost standing on the summit of the apple. Medium width. Slightly ribbed and puckered.
Tube Funnel-shaped.
Stamens Median.
Core line Basal, clasping or median towards basal.
Core Median. Axile, closed or open.
Cells Round.
Seeds Obtuse or acute. Round, plump and straight.
Flesh White. Very firm and crisp. Juicy. Fine-textured.
Aroma Very slight.
Flowers Pollination group 3.
Leaves Medium size. Long acute to long narrow acute. Serrate or bluntly serrate. Medium thick. Flat or very slightly undulating. Slightly upward-folding. Mid green. Undersides fairly downy.

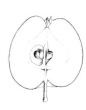

— ASHMEAD'S KERNEL —

Season   December to February
Picking time   Mid October.

This apple was raised in Gloucester, England, by Dr Ashmead in about 1700 and the original tree was still in existence during the early 1800s in Dr Ashmead's garden. However, it was destroyed when the ground was re-allocated for building. It appears that Ashmead's Kernel was well known only in West Gloucestershire until the late 1700s. It is considered to be one of the highest quality late dessert apples available. It received the Award of Merit from the RHS in 1969 and a First Class Certificate in 1981. The flavour is sweet yet a little acid and highly aromatic. It is juicy and refreshing. The trees have moderate vigour, are upright-spreading and spur-bearers, producing spurs freely. The cropping however is erratic. Trees are widely available from fruit nurseries. Trees show good resistance to scab.

Size Medium, 64 × 54mm (2½ × 2⅛″). Slightly larger on a young tree.
Shape Flat-round. Frequently lop-sided. Slight trace of some well rounded ribs. Can be flat-sided. Well flattened at base and apex. Fairly regular.
Skin Yellowish-green (145A) to pale quite bright greenish-yellow (154C). Sometimes partly flushed with ochre-orange (163A) to orange-red (169B). Three quarters to almost completely covered with fine, slightly scaly cinnamon-brown to grey-brown russet (between 164B and 199C), occasionally with some dark-brown russet (200D) round eye. Lenticels not very conspicuous grey-brown or yellow-grey russet dots. Skin very dry.
Stalk Short (7–12mm) within cavity or medium and beyond. Medium to fairly stout (3–3.5mm).
Cavity Medium width and depth. Lined with ochre russet and overlaid with specks of medium to deep brown scaly russet, which radiates out.
Eye Medium. Slightly to half open. Sepals erect convergent, tapering with tips reflexed.
Basin Medium to fairly shallow. Medium width. Regular. Lined with fine cinnamon russet and scaly brown russet. Ribbed and puckered.
Tube Cone-shaped.
Stamens Marginal.
Core line Rather indistinct, almost basal.
Core Median. Axile.
Cells Obovate.
Seeds Acuminate or acute. Fairly plump.
Flesh Yellowish-white tinged very slightly green. Firm, crisp and juicy. Fine-textured.
Aroma Almost nil.
Flowers Pollination group 4.
Leaves Medium size. Broadly acute. Serrate. Medium thick. Flat not undulating. Slightly upward-folding. Mid green. Fairly downy.

— BROWNLEES' RUSSET —

Season   December to March
Picking time   Mid October

This apple was raised and introduced by Mr William Brownlees, a nurseryman in Hemel Hempstead, Hertfordshire, England, in about 1848. It is possibly grown on a small scale in order to supply the limited demand for russets, principally in farm shops. It is fairly widely listed by nurserymen. The fruit has a brisk flavour with a nice balance of sugar and acid and a pleasant nutty taste. The trees are moderately vigorous, upright-spreading and produce spurs very freely: the cropping is moderately good.

Size Medium large, 70 × 58mm (2¾ × 2¼″).
Shape Flat-round to slightly conical. Fairly distinct well rounded ribs. Flattened to rounded at base with some ribs rather pronounced on shoulder. Frequently lop-sided. Rather irregular.
Skin Fairly bright yellowish-green (144B) to greenish-yellow (150B) Some fruits up to quarter flushed with dull orange-brown (168A) to reddish-brown (172A). Skin almost completely covered with greenish-ochre (152B) and fine grey-brown russet (199B) which in parts is overlaid with a fine silvery scaly sheen. A few lenticels are conspicuous as pale yellow russet dots which don't feel raised. The lenticels enter cavity and are larger towards base. No stripes. Skin dry.
Stalk Fairly slender to medium thick (2.5–3mm). Short (8–12mm). Usually level with base or within cavity. Can have fleshy lump on one side.
Cavity Medium width and depth. Sometimes lipped otherwise regular. Usually completely lined with ochre-brown russet. Some have an underlying dark green skin that radiates out from the cavity and over the shoulder.
Eye Smallish medium in size. Closed or partly open. Sepals erect convergent with tips reflexed or broken off. Slightly downy.
Basin Shallow and fairly narrow. Eye sits almost on top of the fruit. Slightly ribbed. Green skin next to the eye, otherwise completely russetted.
Tube Narrow funnel-shaped. (Hogg in *The Fruit Manual* found it to be cone-shaped).
Stamens Median or almost marginal.
Core line Rather indistinct. Median.
Core Median. Axile, closed or open.
Cells Ovate narrowing to a point beneath eye. Tufted.
Seeds Acuminate or acute. Plump. Dark brown.
Flesh Greenish white. Firm. Fine.
Aroma Almost nil.
Flowers Pollination group 3.
Leaves Small. Long acute to narrow acute. Bluntly serrate. Medium thick. Flat not undulating. Dark blue-green. Undersides moderately downy.

114

CRAWLEY BEAUTY

Season   December to March
Picking time   Mid October

This apple was discovered by Mr Cheal, a nursery-man of Crawley in Sussex. He found it in a cottage garden in Sussex in about 1870. He introduced it in 1906 and it was much grown in parts of Sussex during the earlier part of this century. The origin of this apple is not certain: one theory is that it is an American cultivar named Goldhanger, another is that it is French. In the National Fruit Trials it appeared identical with the French cultivar Nouvelle France. It is a useful very late culinary variety to grow in colder areas because of its very late flowering and its hardiness. The trees are moderately vigorous, spreading and produce spurs freely: the cropping is good and the flavour of the fruit is fair. Trees are available from several specialist nurseries.

Size Medium small, 58 × 45mm (2¼ × 1¾″).
Shape Flat-round to round. Distinctly flattened at base and apex. Usually fairly symmetrical but occasionally slightly lop-sided. No ribs. Regular.
Skin Fairly bright yellow-green (144A–144B) becoming pale yellow (154D). Quarter to three quarters flushed with brownish-red (175B) with a crimson hue (185A) nearest the sun, and greenish-ochre (152B) on shaded side. Stripes are fairly distinct brownish-crimson (185A) except on a highly coloured flush where they blend with the flush. Lenticels fairly conspicuous whitish dots. A small amount of thin scarf skin at base or in cavity. Usually no russet. Skin very smooth, shiny and greasy.
Stalk Medium thick, (3mm). Medium length (15–20mm). Protrudes well beyond base.
Cavity Medium width and depth. Regular. There is usually some fine golden-brown russet within.
Eye Medium. Half to fully open. Sepals short, broad based, separated when fully open, erect convergent with some tips reflexed. Fairly downy.
Basin Wide. Medium depth. Usually regular with no ribs. Occasionally some slight puckering. There can be a small amount of cinnamon russet.
Tube Slightly funnel-shaped.
Stamens Median.
Core line Median or towards basal.
Core Median. Abaxile.
Cells Obovate.
Seeds Fairly large. Obtuse. Broad.
Flesh Greenish-white. Firm. Slightly coarse-textured. Not very juicy.
Aroma Nil.
Flowers Pollination group 7.
Leaves Small. Acute. Serrate. Medium to rather thin. Flat not undulating. Very slightly upward-folding. Mid green. Undersides very downy.

FIESTA

Season   Late October to January
Picking time   Mid September

This is a new introduction from East Malling Research Station in Kent, previously known as T31/31. It was selected from a cross between Cox's Orange Pippin and Idared. It is an attractive brightly coloured apple with a similar flavour to a Cox, and is proving to be of great commercial interest because of its consistently good cropping habit and good storage properties. It received a Preliminary Commendation from the RHS in 1987. Experiments suggest that Fiesta shows a degree of self fertility and that the flowers withstand low temperatures well. The trees are moderately vigorous, upright-spreading and spur-bearers . The fruit has quite a rich vinous flavour and is quite sweet with a nice tang, making it refreshing.

Size Medium small, 58 × 51mm (2¼ × 2″) to medium, 67 × 51mm (2⅝ × 2″).
Shape Round to flat-round. Usually slightly lop-sided. No ribs or may be a slight trace of one or two very broad well-rounded ones. Regular.
Skin Greenish-yellow (150B) becoming yellow (3B). Half to three quarters covered with bright red flush which can be a very dense crimson-red (46A) on the sunny side or a thinner and more speckled red (46B) merging into the yellow skin. Short broken stripes of crimson-red (46A) to purplish-crimson (187B) on dense flush. Lenticels fairly inconspicuous whitish dots. Occasional small ochre or grey-brown russet patches. Skin smooth, becoming greasy.
Stalk Fairly slender to medium stout (2.5–3.5mm). Quite long to long (22–30mm). Frequently set at an angle.
Cavity Fairly wide. Medium depth. Frequently lipped. There can be some scaly brown russet.
Eye Medium size. Closed or slightly open. Sepals fairly short, erect or rather flattish convergent with tips reflexed or broken off.
Basin Medium width and fairly shallow. Slightly puckered. Very downy. There is sometimes a small amount of brown russet.
Tube Funnel-shaped.
Stamens Marginal.
Core line Faint. Median towards marginal.
Core Median. Axile.
Cells Round or roundish obovate. Slightly tufted.
Seeds Large. Obtuse. Flat on one side. Curved.
Flesh Yellowish. Rather coarse-textured. Firm and crisp. Juicy.
Aroma Very slight, rather acid.
Flowers Pollination group 3.
Leaves Medium to small. Acute to narrow acute. Serrate. Medium thick. Flat or undulating. Upward-folding. Dark blue-green. Downy.

─── NONPAREIL ───

Season   December to March
Picking time   Mid October

This old and highly prized dessert apple has been in existence in England for centuries. It is generally thought that it was brought here from France by a Jesuit in Queen Mary's or Queen Elizabeth I's time. It is primarily a garden apple for the connoisseur. The word Nonpareil, meaning a person or thing that is unsurpassed, was applied to a number of apples considered to be of high quality. The flavour is good with a nice balance of sweetness and acidity. The cropping is light to moderate. The trees are rather weak in growth, spreading and produce spurs very freely. Trees do not appear to be listed by nurserymen.

Size Medium small, 58 × 48mm (2¼ × 1⅞").
Shape Flat-round rather conical. Very slight trace of one or two ribs on some fruits. Base flattened at centre, rounded at shoulders. Flattened at apex. Symmetrical or lop-sided. Regular.
Skin Pale dull yellowish-green (145A) to pale greenish yellow (150C). Some fruits slightly flushed with greenish-ochre (153C) and blotched with pale reddish-brown (173A). No stripes. Partly covered with very fine pale ochre and grey-brown russet. There is a pronounced area of pale grey-brown russet covering apex. Lenticels inconspicuous on cheeks as very pale brown russet dots, but with some fruit they are surrounded by a prominent patch of greyed-purple (183A). Lenticels fairly noticeable at base as whitish dots which also enter cavity. Skin slightly rough and dry.
Stalk Fairly slender to medium (2.5–3mm). Long (18–28mm). Extends well beyond base.
Cavity Medium width. Medium depth. Regular. Usually lined with fine light brownish-grey russet which spreads and scatters over base.
Eye Medium size. Half to completely open. Sepals broad based tapering to a fine point or broken off, erect or convergent but not touching, with tips reflexed. Slightly downy.
Basin Very shallow. Medium width. Slightly ribbed. Mostly lined with fine grey-brown russet.
Tube Cone or funnel shaped.
Stamens Marginal or towards median.
Core line Almost basal.
Core Median. Axile.
Cells Ovate or roundish.
Seeds Obtuse. Fairly plump. Slightly angular.
Flesh Firm. Fine-textured. Greenish-white. Juicy.
Aroma Nil.
Flowers Pollination group 3.
Leaves Smallish. Acute. Bluntly serrate. Medium thick. Flat not undulating. Slightly upward-folding. Mid greyish-green. Slightly downy.

─── KING'S ACRE PIPPIN ───

Season   December to March
Picking time   Mid October

It is stated that this very late dessert apple was introduced in 1899 by King's Acre Nurseries but the exact origin is not specified. It is said to be a Sturmer Pippin and Ribston Pippin cross, but which direction is not recorded. It received an Award of Merit from the RHS in 1897. It has always been highly regarded for flavour, although its appearance leaves something to be desired. The flavour is richly aromatic and refreshing with a slight acidity. The cropping is slow to begin with when the trees are young but improves when the trees mature. This variety is not suitable for the small garden as it has vigorous growth and makes a large spreading tree. It is a partial tip-bearer. Trees are available from a few nurseries.

Size Medium large, 73 × 67mm (2⅞ × 2⅝").
Shape Round-conical, sometimes rather flat-sided and squarish. Can be symmetrical but usually lop-sided. Fairly distinct well rounded ribs with one or two sometimes more pronounced than the others. Irregular. Can be slightly five-crowned at apex.
Skin Dull green (146C) becoming dull greenish-yellow (151A–151D). Fruits either not flushed, slightly flushed with a dull ochre-green (152B), to half flushed with brownish-red (178B). Some short, broken stripes of dull deep red (46A). All fruits have patches or dusting of fine grey-brown russet, particularly at apex where it is usually a larger more solid area. Lenticels are conspicuous pale grey-brown russet dots. There is usually some rather thin patchy scarf skin at base and sometimes on cheeks. Skin fairly smooth and dry.
Stalk Fairly stout (3.5mm). Medium length (17mm). Extends beyond base.
Cavity Medium to wide. Medium depth. Often lipped. Usually lined with fine grey-brown russet.
Eye Medium size. Closed or partly open. Sepals broad based, erect convergent sometimes not quite touching, with tips reflexed. Fairly downy.
Basin Medium width and fairly shallow. Ribbed and puckered. Russetted with grey-brown russet.
Tube Funnel-shaped.
Stamens Median, above core line.
Core line Median or almost basal. Rather faint.
Core Median. Axile.
Cells Obovate. Slightly tufted.
Seeds Fairly large. Dark brown. Fairly plump. Acuminate. Slightly curved.
Flesh Greenish-yellow. Coarse-textured. Firm and crisp. Juicy.
Aroma Slight rich aromatic scent.
Flowers Pollination group 4.
Leaves Medium size. Acute to broadly acute. Sharply serrate. Thin. Flat. Dark green. Slightly downy.

LORD HINDLIP

This very late dessert apple was first exhibited to the fruit committee of the RHS by Mr Watkins of Pomona Farm, Hereford in 1896. It was given an Award of Merit. In 1898, it received a First Class Certificate when shown by Mr P.C.M. Veitch. The fruit is a very late dessert type, with rich and distinctive vinous flavour. The trees have moderate vigour, are upright-spreading in habit and produce spurs fairly freely. The cropping is good. Trees are available from one or two specialist nurseries.

Size Medium large, 73 × 73mm (2⅞ × 2⅞"), or 64 × 77mm (2½ × 3").

Shape Conical, to long-conical. Very broad base sloping to narrow apex. Base flattened, or rounded. Often a little lop-sided. Fairly distinct ribs often with one slightly larger. Can be rather angular and flat-sided towards apex. Five-crowned at apex.

Skin Pale greenish-yellow (151D) to pale yellow (10A). Half to three quarters flushed either with brownish-orange (169A) speckled and dotted with red (45A), to crimson-red (46A) on well coloured fruits. Some short broken rather broad stripes of crimson-red (46A). Lenticels numerous and distinct greenish-grey russet dots. Variable amounts of ochre-grey or grey-brown russet, sometimes netted. Skin dry.

Stalk Medium thick (3mm). Fairly short to medium (15–20mm). Level with base or slightly beyond.

Cavity Wide and deep. Regular. Lined with fine dirty ochre and scaly dark grey-brown russet which streaks and scatters over base.

Eye Rather small. Closed or very slightly open. Sepals erect convergent or slightly connivent and rather pressed together, with tips reflexed or frequently broken off. Fairly downy.

Basin Narrow. Medium depth. Slightly ribbed. Patched or netted with fine grey-brown or pale grey russet. Can be slightly beaded.

Tube Deep narrow cone or deep funnel.

Stamens Median or almost marginal.

Core line Rather faint. Median.

Core Distant to very distant. Axile, open.

Cells Ovate. Sometimes long and narrow. Slightly tufted.

Seeds Acuminate or acute. Plump. Straight.

Flesh White sometimes tinged pink under the skin. Rather coarse-textured. Firm and juicy.

Aroma Very slight.

Flowers Pollination group 3.

Leaves Small. Narrow acute. Serrate or bluntly serrate. Medium thick. Very upward-folding. Light grey-green. Undersides very downy.

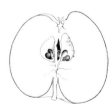

LANE'S PRINCE ALBERT

This late culinary apple is thought to have been raised by Thomas Squire of Berkhamsted in about 1840 from Russet Nonpareil × Dumelow's Seedling. It was exhibited by H. Lane & Son of Berkhamsted in 1857 and in 1872 received a First Class Certificate from the RHS. The apple was named to celebrate a visit to Berkhamsted by Her Majesty Queen Victoria and Prince Albert. In the past, it has been grown commercially but its popularity declined in the 1940s. It is available from fruit tree nurserymen today and there are still a few commercial plantings. The flavour of the fruit is fair, acidic and rather watery, and it stays fairly intact on cooking. The trees are hardy and suitable for growing in the North but are susceptible to mildew. They are suitable for the small garden being of weak growth and rather dwarfish, upright then spreading in habit. They produce spurs freely and the cropping is heavy.

Size Large, 77 × 67mm (3 × 2⅝").

Shape Round-conical. Often slightly lop-sided. Rather indistinct broad ribs. Slightly five-crowned at apex. Sometimes ribs slightly prominent at base otherwise base flat. Slightly irregular.

Skin Bright grass-green (144B) changing to light yellow (6C). Some fruits slightly to one third flushed with greyed red (178C) with broad broken stripes of a deeper greyed red (178B). Some fruits barely flushed but just greyed red stripes over a slightly orange skin (164A). Lenticels fairly conspicuous pinky-grey russet or pale ochre dots. Skin very smooth and shiny. No russet.

Stalk Fairly slender (2.5mm). Short to fairly short (10–17mm). Within cavity or protrudes beyond.

Cavity Fairly wide and deep. Can be partly lined with fine grey-brown russet. Dark green, partly lined with fine grey-brown russet and some scarf skin which can extend on to base. Some lenticels in cavity appearing as large oval whitish dots.

Eye Smallish. Closed or slightly open. Sepals small, connivent with tips reflexed. Very downy.

Basin Fairly deep. Medium width. Ribbed.

Tube Cone or funnel-shaped.

Stamens Median.

Core line Basal, clasping.

Core Median. Abaxile.

Cells Elliptical or ovate. Tufted.

Seeds Numerous. Acuminate. Plump. Straight.

Flesh Greenish-white. Fine-textured. Firm. Juicy.

Aroma Very slight rather acid scent

Flowers Pollination group 3.

Leaves Large. Acute. Broadly serrate. Medium thick. Mostly flat. Mid green. Slightly downy.

LORD BURGHLEY

Season   January to April
Picking time   Mid October

This very late dessert apple was raised as a seedling in the gardens of the Marquis of Exeter at Burghley Park near Stamford. It first fruited in 1834. It was introduced in 1865 by Mr Matheson, the gardener at Burghley Park and was distributed in that year by Mr House of Peterborough. It received a First Class Certificate from the RHS in 1865. The fruit has a rich aromatic flavour, sweet with a nice balance of acidity. The cropping is fairly good. The trees are moderately vigorous, upright spreading and have burrs, which are small rough outgrowths or burls on the trunk and lower branches from which leaf shoots sprout. It is a spur-bearer and produces spurs fairly freely. Trees are listed by a few nurseries.

Size Small to medium, 67 × 61mm (2⅝ × 2⅜″).
Shape Round, slightly conical. Slightly flattened at base and apex. Fairly prominent ribs which can be rather angular and are particularly noticeable towards apex. Slightly five-crowned at apex. Slightly irregular.
Skin Pale yellowish-green (151A) becoming pale yellow (6C). Quarter to three quarters flushed brownish-orange (171B) with indistinct stripes and speckles of a paler orange-red (34A). Well coloured fruits flushed with orange-red (34A) with indistinct rather thin, broken stripes of crimson-red (46A). Variably patched and dusted with grey-brown russet which looks rather silvery when on a bright flush. Lenticels very noticeable, numerous, large, grey or dark brown russet dots. Skin dry and fairly smooth.
Stalk Stout (4mm). Medium length (15–20mm). Protrudes beyond base.
Cavity Medium width. Rather shallow. Sometimes lipped with stalk set at an angle. Partly or completely lined with fine greyish-green russet.
Eye Medium size. Partly open. Sepals erect convergent, long and narrow with tips reflexed. Downy.
Basin Medium width, medium depth. Clearly ribbed.
Tube Deep cone, sometimes almost funnel-shaped.
Stamens Median sometimes marginal.
Core line Median slightly towards basal.
Core Median. Axile.
Cells Obovate
Seeds Obtuse. Medium plump. Rather angular and sometimes flat-sided. Dark brown. Fairly wide.
Flesh Yellowish-white. Firm but tender. Rather coarse-textured. Juicy.
Aroma Moderately strong, sweetly scented.
Flowers Pollination group 4.
Leaves Medium size. Oval. Bluntly serrate. Medium to fairly thick. Flat with some very slightly upward-folding. Mid green. Very downy.

DUKE OF DEVONSHIRE

Season   January to March
Picking time   Early October

This well known late to very late dessert apple was raised in England in 1835 at Holker Hall in Lancashire, home of the Duke of Devonshire. It was raised by the gardener at the Hall, Mr Wilson, who named the apple after his employer. It was introduced in about 1875. Duke of Devonshire used to be grown on a small scale for the markets earlier this century. This apple is still listed by a few specialist nurseries. It is resistant to scab and therefore suitable for growing in the West Country. The trees are moderately vigorous, spreading and spur-bearers. The cropping is good. The fruit is rich in flavour, juicy and slightly acidic, making it refreshing.

Size Medium, 61 × 54mm (2⅜ × 2⅛″).
Shape Flat-round to round-conical. Slight trace of ribs. Can be quite flattened at base and apex or rather rounded at base. Symmetrical or slightly lop-sided. Regular.
Skin Rather dull green (144C nearest but should be more yellow) becoming pale greenish-yellow (151D) and eventually yellow (7A). Partly covered with patchy and netted fine grey-brown russet (199A) usually concentrated around the apex and spreading over cheeks. Lenticels conspicuous large grey russet dots appearing slightly raised. Skin very dry and slightly textured.
Stalk Stout to very stout (4–5mm). Very short (5mm) well embedded within cavity.
Cavity Narrow and shallow. Skin green or ochre lined with fine grey-brown russet, sometimes finely scaled, which streaks and scatters over base.
Eye Medium size. Slightly to half open. Sepals broad based, short and rather flat or slightly erect convergent with tips reflexed. Fairly downy.
Basin Shallow. Medium width. Usually fairly distinctly ribbed and puckered. Lined with grey-brown russet with some green skin usually showing just around the eye.
Tube Cone-shaped.
Stamens Median towards marginal.
Core line Median or basal sometimes appearing to pull out the sides of the tube and immediately diverging outwards towards the sides of the apple.
Core Median. Axile.
Cells Obovate.
Seeds Fairly numerous. Fairly plump. Acute.
Flesh Creamy-white tinged slightly green. Fairly fine texture. Firm and hard, fairly juicy.
Aroma Almost nil, slightly woody.
Flowers Pollination group 4.
Leaves Medium size. Acute. Bluntly serrate. Medium thick. Slightly undulating and slightly upward-folding. Mid greyish-green. Very downy.

122

D'ARCY SPICE

Season   December to April
Picking time   Late October to early November

This old dessert apple was found in the gardens of The Hall, Tolleshunt d'Arcy, near Colchester, Essex in about 1785, where it is said many trees of this variety existed at that time. It was always known as D'Arcy Spice or Spice Apple until 1848 when Mr John Harris, a nurseryman at Broomfield near Chelmsford, sold it under the name of Baddow Pippin. This apple was not grown much outside East Anglia but it is listed by several nurserymen today. It requires a hot dry summer to gain the spicy flavour for which it was named. The fruit is richly aromatic, sweet yet acid with a nice balance, but the skin is tough. The cropping is erratic and the fruits shrivel easily unless properly stored. The trees are vigorous, upright-spreading and partial tip-bearers.

Size Medium, 67 × 58mm (2⅝ × 2¼″).
Shape Oblong. Flattened or rounded at base, flattened at apex. Often slightly lop-sided. Fairly distinct well rounded ribs, frequently with one larger. Five-crowned at apex. Irregular.
Skin Light yellowish-green (144B) becoming pale greenish-yellow (150C). There can be a slight pinkish-brown (174B) to purplish-brown (174A) flush. Variably covered with finely scaled cinnamon (164A) or grey-brown (165A) russet. A scruffy looking apple. Lenticels conspicuous slightly raised grey-brown or cinnamon russet dots. Skin very dry.
Stalk Medium to stout (3–4mm). Short, (12mm). Embedded well and sometimes fills cavity.
Cavity Fairly narrow and fairly deep. Skin lighter green, with an almost translucent golden hue. Lined with ochre sometimes scaly russet.
Eye Quite large. Partly open. Sepals geen at base, broad based, rather long, convergent with half or just the tip reflexed. Fairly downy.
Basin Medium width and depth. Ribbed. Partly or completely lined with finely scaled cinnamon-russet which can run concentrically round basin.
Tube Wide. Cone-shaped.
Stamens Median or towards basal.
Core line Towards basal or basal, clasping. Follows close to outline of cell. There can be a second marginal core line.
Core Median. Axile.
Cells Roundish ovate.
Seeds Acute. Quite large. Flattish. Not very plump.
Flesh White tinged green. Firm. Fine. Juicy.
Aroma Very slight.
Flowers Pollination group 4.
Leaves Medium size. Acute. Serrate. Medium to rather thin. Flat. Some downward-folding. Mid green. Slightly downward-hanging. Downy.

EDWARD VII

Season   December to April
Picking time   Mid October

A very late culinary apple that was introduced in 1908 by Messrs Rowe of Worcester, England. The parents are said to be Blenheim Orange × Golden Noble. It was first recorded in 1902 and received the Award of Merit from the RHS in 1903. It has always been planted on a limited scale commercially in the United Kingdom and is also a useful garden apple because of its neat upright growth. Trees are rather slow to come into bearing and they are only moderate croppers. They flower late and therefore sometimes escape late spring frosts, but do not always set well due to lack of pollinators. The trees are hardy and scab resistant. They are spur-bearers. The fruit is acid with good flavour and cooks to a somewhat red translucent puree.

Size Large, 83 × 70mm (3¼ × 2¾″).
Shape Round to flat-round. Fairly symmetrical but can be slightly lop-sided. Usually no ribs but can be a very slight trace. Regular.
Skin Fairly bright green (144B) becoming light yellow-green (154B) eventually pale yellow (7C). Occasionally a slight mottled pale pinkish-brown flush (174B). Some patchy scarf skin at or towards base and surrounding some lenticels. Often a raised green or dark grey-brown hair line, sometimes two. Lenticels conspicuous grey-brown or whitish russet dots, also some green dots which lie beneath the skin. Skin smooth and dry.
Stalk Very stout (5mm). Short (6mm). Sometimes fleshy. Usually level with base.
Cavity Medium width. Very shallow. Sometimes lipped. Lenticels enter cavity as largish round or oval whitish dots. There can be some fine grey-brown russet within but usually russet free.
Eye Quite large. Open. Sepals broad based, sometimes separated at base. Erect convergent with tips reflexed. Slightly downy.
Basin Medium width. Shallow. Regular. Can be very slightly ribbed and puckered.
Tube Deep funnel-shape.
Stamens Median at neck of funnel.
Core line Median.
Core Median. Axile, sometimes open.
Cells Obovate or elliptical.
Seeds Acuminate. Fairly plump. Sometimes rather angular. Regular or slightly curved.
Flesh Creamy white. Rather coarse-textured. Fairly juicy. Firm but tender. Skin a little tough.
Aroma Nil. Fairly sweetly aromatic on storing.
Flowers Pollination group 6.
Leaves Medium size. Broadly oval to broadly acute. Serrate or bluntly serrate. Medium thick. Slightly undulating. Very dark grey-green. Very downy.

TYDEMAN'S SLATE ORANGE

Season   December to April
Picking time   Mid October

A late to very late dessert apple which was raised in 1930 at the East Malling Research Station at Maidstone in Kent by Mr H.M. Tydeman from Laxton's Superb × Cox's Orange Pippin. It was introduced in 1949 and has been grown commercially on a small scale. It received the Award of Merit from the RHS in 1965. The fruit has a rich Cox-like flavour with a nice balance of sugar and acid. The skin is slightly tough. The cropping is good but tends to over-crop with small fruits. Fruits should be stored in polythene bags as they tend to shrivel. The trees are vigorous, upright then spreading with long laterals. They produce abundant new growth and spur freely. This variety is widely listed by fruit tree nurseries today.

Size Medium small, 58 × 54mm (2¼ × 2⅛").
Shape Conical. Usually symmetrical. Almost no trace of ribs. Flattened at base narrowing to a slightly flattened apex. Regular.
Skin Pale yellowish-green (144C) becoming dull greenish-yellow (151D). Quarter to three quarters flushed with dull brownish-purple (178B). Indistinct, broken stripes of slightly deeper brownish-purple (178A). Patched and sometimes slightly netted with fine greyish-ochre and brownish-grey russet. Some scarf skin chiefly at or towards base. Lenticels inconspicuous pale or dark grey-brown russet dots. Skin dry and slightly rough.
Stalk Fairly slender to fairly stout (2.5–3.5mm). Shortish to medium (15–22mm). Protrudes slightly or well beyond cavity.
Cavity Medium width, fairly shallow. Sometimes lipped, otherwise regular. Skin ochre green lined with scaly grey-brown russet, extends over base.
Eye Large. Open. Sepals broad and long, often slightly separated at base, erect with tips slightly reflexed. Stamens present. Almost no down.
Basin Rather shallow and medium width. Regular. Very slightly puckered. Partly or completely lined with fine pale brown russet.
Tube Cone-shaped.
Stamens Median.
Core line Basal, clasping.
Core Median. Axile, open.
Cells Roundish ovate.
Seeds Acuminate to acute. Plump. Regular or curved.
Flesh Creamy yellow. Firm. Fine-textured. Juicy.
Aroma Almost nil. Slightly woody.
Flowers Pollination group 4.
Leaves Medium size. Acute. Bluntly serrate. Medium thick. Flat not undulating. Some slightly upward-folding. Light grey green. Slightly downy.

PIXIE

Season   December to March
Picking time   Mid October

This is a fairly recent introduction. It is a high quality, rather small, late dessert apple which was raised in England in 1947 at the National Fruit Trials. The parentage does not appear to be recorded, but it is thought it may have been a seedling from Cox's Orange Pippin or Sunset. It received the Award of Merit from the RHS in 1970 and a First Class Certificate in 1972. It is mainly a garden variety, being too small for commercial use, but could well have a use for farm shops. The fruit is delicious; crisp, juicy, sweet yet refreshing, attractive to look at and not too large. The trees are moderately vigorous, wide-spreading and spur-bearers. The cropping is good. This apple is listed by a few specialist nurseries.

Size Medium, 64 × 51mm (2½ × 2").
Shape Flat-round. Flattened at base and apex. No ribs. Usually symmetrical but can be slightly lopsided. Regular.
Skin Greenish-yellow (2B). Quarter to three quarters flushed with orangey-red (34A), mottled and dotted towards the edges. Short broken stripes of red (46B). Some greenish-ochre or greenish-grey russet dots and patches. Lenticels fairly conspicuous slightly raised pale ochre-grey or grey-brown dots. Sometimes a small amount of patchy scarf skin at base. Skin smooth and dry and slightly bumpy.
Stalk Slender to medium (2–3mm). Medium length (20mm). Sometimes fleshy.
Cavity Medium width and medium to fairly deep. Regular. Some greenish-ochre sometimes slightly scaly russet within cavity which sometimes streaks or scatters out over shoulder.
Eye Small. Closed or slightly open. Sepals short or long, convergent with half to three quarters reflexed. Downy. Stamens sometimes present.
Basin Medium width and fairly shallow. Slightly ribbed, occasionally beaded. There can be some flecks of fine grey russet.
Tube Small funnel-shaped.
Stamens Median.
Core line Basal, clasping.
Core Median. Axile.
Cells Obovate.
Seeds Acute. Very large for size of apple. Dark brown. Numerous. Straight. Fairly plump.
Flesh Creamy white. Firm. Fine-textured and juicy.
Aroma Nil. Very slightly aromatic when cut.
Flowers Pollination group 4.
Leaves Medium size. Acute to rather long acute. Bluntly serrate. Thin. Flat or slightly undulating. Upward-folding. Light yellowish-green. Downy.

Season   December to April
Picking time   Mid October

A late to very late culinary apple, which was raised by Charles Ross at Welford Park, Newbury, Berkshire from Warner's King × Northern Greening. It was first recorded in 1906 when it received the Award of Merit from the RHS. It was introduced in 1908 by Messrs Cheal of Crawley and was awarded a First Class Certificate. It is grown on a small scale commercially in the United Kingdom but trees have a very limited availability from nurseries. It is a good variety for frosty areas due to its late flowering and it shows some resistance to scab. The trees are moderately vigorous, upright-spreading and spur-bearers: the cropping is moderate. The fruit is sub-acid. The slices remain intact when cooked and the flesh had a good though not strong flavour.

Size Large to very large, 83 × 73mm (3¼ × 2⅞″) to 92 × 77mm (3⅝ × 3″).
Shape Round to oblong. Much flattened at base and apex. Slightly ribbed with broad or rather angular ribs. Slightly five-crowned at apex. Can be a little flat-sided. Occasionally lop-sided. Slightly irregular.
Skin Bright yellowish-green (between 144B and 150B), to greenish-yellow (151D). Indistinctly striped and patched with green under the yellow. Can be slightly to half flushed with brownish-crimson (185A) or a paler greyed brown (177B). Short broken stripes of brownish-crimson (178A). Variable amounts of scarf skin at base. Lenticels fairly distinct whitish or grey-brown russet dots. Usually russet free or slight fine grey-brown russet.
Stalk Stout (4mm) and shortish (10–15mm). Set well within cavity. Sometimes fleshy.
Cavity Medium width and medium depth. Can be partly lined with fine grey-brown russet.
Eye Large. Closed or slightly open. Sepals broad, erect convergent with tips well reflexed. Downy.
Basin Wide and fairly deep. Ribbed and puckered, sometimes beaded.
Tube Slightly or definitely funnel-shaped. Deep.
Stamens Median.
Core line Median.
Core Median. Abaxile.
Cells Ovate long, somewhat lanceolate. Tufted.
Seeds Acute. Numerous. Small.
Flesh Creamy white with slightly greenish tinge. Rather coarse-textured. Soft but firm.
Aroma Slightly acid before and after cutting.
Flowers Pollination group 4.
Leaves Medium size. Acute. Sharply serrate. Medium thick. Flat not undulating. Mid to rather light yellowish-green. Undersides very downy.

Season   December to April.
Picking time   Mid October

It is suggested that this apple may have originated in Roman times and that it probably came from Europe as it was extensively grown throughout France and Germany. The first record of it is in about 1613. There are a vast number of synonyms attached to this apple, several of which came from the family residences where this apple was grown, for example Garnon's Pippin, home of Sir John Cotterell, who it is thought may have introduced it to England. The name is derived from Corps Pendu, referring to the shortness of stalk. Court Pendu Plat is still listed by a few specialist nurserymen today and is a useful variety to grow in areas prone to spring frosts due to its late flowering habit and general hardiness. The tree is of rather weak to moderate growth and makes a smallish upright-spreading tree suitable for the smaller garden. It is a spur-bearer, producing spurs freely and the cropping is good. The fruit has a rich, aromatic flavour with a good balance of acid and sugar.

Size Medium, 61 × 45mm (2⅜ × 1¾″).
Shape Flat. Very flattened at base and apex. Faint trace of ribs. Symmetrical. Regular.
Skin Greenish-yellow (150B) becoming yellow (7C). Quarter to three quarters flushed with orange (169A) to orange-red (34A) deepening to scarlet-red (42A) on well coloured fruits. Indistinct short broken stripes of crimson-red (46A) slightly more obvious on the orange flushed fruits. Lenticels conspicuous and numerous large greenish-ochre russet dots. Some thin patches and dusting of greyish-russet. Skin dry.
Stalk Fairly slender to medium (2.5–3mm). Short (10mm). Usually within cavity.
Cavity Medium width. Fairly deep. Regular. Lined with greyish-ochre russet with fine light brown scaling, which may streak over base.
Eye Large. Open. Sepals short, erect convergent with tips reflexed. Stamens present. Fairly downy.
Basin Wide and fairly deep. Slightly ribbed. Regular. Some ochre or grey-brown russet dots and dashes lying concentrically round basin.
Tube Wide funnel-shaped, or almost cone-shaped.
Stamens Median.
Core line Basal or median towards basal.
Core Median. Axile.
Cells Rather small. Obovate.
Seeds Obtuse. Large, broad, plump or rather flat.
Flesh Cream. Firm, fine-textured. Fairly juicy.
Aroma Slight, sweetly aromatic. Stronger when cut.
Flowers Pollination group 6.
Leaves Medium to smallish. Acute. Broadly, sharply serrate. Medium to rather thin. Upward-folding. Mid greyish-green. Undersides downy.

Season   December to April
Picking time   Mid October

A high quality, late to very late dessert apple which was raised in England in 1920 at Welford in Berkshire by William Pope from Cox's Orange Pippin × Worcester Pearmain. It was introduced in 1935 as Winter King when it received the Award of Merit from the RHS. In 1944 the name was changed to Winston and in 1951 it received another Award of Merit under that name. It is grown on a small scale commercially in the UK and is listed by several nurserymen. The fruit has a good aromatic flavour, sweet and slightly acid. The skin is rather tough. The cropping is heavy and tends to overcrop and produce small fruits. The trees are moderately vigorous, upright-spreading, spur-bearers, and suitable for growing in the North and West.

Size Medium small, 57 × 54mm (2¼ × 2⅛″).

Shape Round-conical to oblong-conical. Usually symmetrical but can be a little lop-sided. Only a very slight trace of ribs. Flattened at apex, some slightly flattened at base, some rounded. Regular.

Skin Rather dull greenish-yellow (151C). Quarter to three quarters flushed with dull brownish-red (178C) or darker purplish-red (178B). Fairly distinct narrow or broad broken stripes of dull brownish-crimson (185A) to purplish-crimson on darker flush (187B). Lenticels quite distinct whitish or grey-brown russet dots. There can be a slight dusting of grey-brown russet and occasional patches of russet. Skin smooth and dry.

Stalk Medium to very stout (3–5mm). Short to medium (10–18mm). Level with base or protrudes beyond.

Cavity Medium to shallow. Medium to narrow. Sometimes lipped. Lined with fine ochre-brown russet which can streak over base.

Eye Medium size. Partly to half open. Sepals long, erect-convergent. Fairly downy.

Basin Medium width and depth. Regular. Slightly ribbed. Some fine grey-brown russet.

Tube Funnel-shaped. Sometimes with lower portion of funnel filled making core line look basal.

Stamens Median sometimes towards marginal.

Core line Median (See tube).

Core Median. Axile.

Cells Small, obovate, sometimes roundish.

Seeds Acute to acuminate. Plump. Fairly straight.

Flesh Cream tinged green, sometimes tinged pink. Fine-textured. Fairly juicy. Skin rather tough.

Aroma Slight, sweetly spicy.

Flowers Pollination group 4.

Leaves Medium size. Acute. Bluntly serrate. Thick and leathery. Flat not undulating. sometimes downward-folding. Dark green. Very downy.

Season   December to April
Picking time   Early October

This very late dessert apple is of American origin. It is recorded that in 1791 Mr George Wheeler brought seeds from Dover, Duchess County, New York, to his new nursery in Yates County. In 1796 Abraham Wagener bought this seedling nursery and by the end of the 1800s Wagener was available from nurseries throughout the United States. It came to England and received the Award of Merit from the RHS in 1910. It is listed by a few specialist nurserymen in the UK. The fruit has a rather weak and insipid flavour but is quite refreshing. The skin is tough. The cropping is heavy with a biennial tendency and it can overcrop with small fruits. It makes a compact tree, moderate of vigour and upright in habit, and it is a spur-bearer. The trees are fairly hardy and suitable for the North and West.

Size Medium large, 70 × 58mm (2¾ × 2¼″).

Shape Flat-round. Some fairly distinct well-rounded ribs. Symmetrical or often lop-sided. Irregular.

Skin Dull pale greenish-yellow (150C–145A). Quarter to three quarters flushed with bright pinkish-red (53B). Rather indistinct short broken stripes of crimson (53A). Mottled and streaked with varying amounts of scarf skin chiefly at base. Areas of thin laced and flecked ochre-brown russet, chiefly at apex but sometimes spreading down cheeks. Lenticels inconspicuous small pale whitish dots. Skin smooth and dry.

Stalk Fairly slender (2.5mm). Medium length (18–24mm). Protrudes beyond base.

Cavity Deep. Variable width: wide, or rather compressed with almost ribbed sides. Some ochre or grey-brown russet which can scatter over base.

Eye Medium small. Closed or partly open. Sepals erect-convergent, sometimes slightly connivent, with tips slightly reflexed. Downy.

Basin Medium width and depth. Irregular. Sometimes ribbed. Some lacy ochre-brown russet. Can be slightly beaded.

Tube Deep funnel-shaped or cone-shaped.

Stamens Marginal.

Core line Sometimes faint. Median towards basal.

Core Median. Axile or abaxile.

Cells Roundish or roundish obovate.

Seeds Obtuse. Small to medium. Slightly curved.

Flesh Creamy white. Crisp and firm. Fine-textured. Juicy.

Aroma Very slight. Sweetly scented.

Flowers Pollination group 3.

Leaves Medium. Acute to broadly acute. Serrate or bluntly serrate. Rather thin. Slightly undulating and upward-folding. Mid yellowish-green. Slightly downy.

STURMER PIPPIN

Season   January to April
Picking time   Mid to end November

This is a fairly old, very late, dessert apple which was raised by a nurseryman named Dillistone at Sturmer near Haverhill in Suffolk, England. It was recorded in the *Gardeners Chronical* of 1847 that a plant of it was presented to the Horticultural Society by Mr Dillistone in 1827 and it was listed as a first rate variety in the Society's Catalogue of Fruits. The parents of this apple are believed to be Ribston Pippin × Nonpareil. It is a high quality apple but requires a warm season to reach its full potential. The fruit has a rich flavour, good and juicy, and the cropping is good. The trees are moderately vigorous, compact, and produce spurs very freely. This cultivar is listed by several fruit tree nurserymen.

Size Medium, 64 × 54mm (2½ × 2⅛″).
Shape Round-conical to oblong conical. Frequently a little lop-sided. Flattened at base and apex. Fairly distinct ribs. Can be a little flat-sided. Slightly irregular. Five-crowned at apex.
Skin Bright green (144B) becoming greenish-yellow (151D). Slightly to half flushed with dull khaki (between 152C and 199A) to purplish-brown (172A). Flush stops rather abruptly and doesn't merge much. Some slight suggestion of stripes (172A). Frequently russetted with fine cinnamon russet around apex and some very small brown russet patches on cheeks. Lenticels distinct on flush as large yellowish areolar dots, especially towards base, otherwise whitish or grey-brown russet dots. Skin smooth and dry.
Stalk Fairly slender to stout (2.5–4mm). Short to fairly long (12–25mm). Extends beyond base.
Cavity Wide and fairly deep. Some greenish-ochre or dark grey-brown russet which can streak and scatter over base.
Eye Fairly small. Closed or slightly open. Sepals broad based and tapering, erect convergent or slightly connivent, with tips well reflexed or broken off. Very downy.
Basin Medium depth and fairly wide. Ribbed. Usually green. Occasional dusting of cinnamon russet.
Tube Slightly funnel-shaped.
Stamens Marginal.
Core line Median towards basal.
Core Slightly sessile. Axile.
Cells Obovate.
Seeds Large. Obtuse. Fairly plump. Straight.
Flesh White tinged green. Firm. Fine-textured. Juicy.
Aroma Almost nil.
Flowers Pollination group 3.
Leaves Medium size. Acute. Bluntly serrate. Medium thick. Flat not undulating. Slightly upward-folding. Mid green. Very downy.

GRANNY SMITH

Season   January to April
Picking time   Mid October

This apple originated in Australia. It was raised quite by chance from a seed thrown out by Mrs Thomas Smith of Ryde in New South Wales. The origin of the seed is thought to be from a French Crab, open pollinated. The tree was known to be fruiting in 1868. This apple arrived in England in 1935. Granny Smith is grown in Australia, South Africa, New Zealand and parts of Europe and North America, but it requires a warm climate in order to develop any sugars or flavour and is only suitable as a cooker when grown in the UK. The flavour is insipid but the flesh is hard and juicy. The skin is a little tough. The trees are of moderate vigour, upright-spreading and spur-bearers. They make small trees and the cropping is good. Granny Smith is listed by fruit tree nurserymen in the UK.

Size Medium, 64 × 61mm (2½ × 2⅜″).
Shape Round-conical. Flattened at base with shoulders rounded. Slightly flattened at apex. Five-crowned at apex. Slightly ribbed. Fairly regular and fairly symmetrical.
Skin Grass green (144A). Fruit can be slightly flushed with purplish-brown (178A) or brownish-ochre (between 199A and 164A), which can appear rather striped, otherwise no stripes. Lenticels very conspicuous numerous large whitish or pink areolar dots. These dots usually enter the cavity. Skin very smooth and dry.
Stalk Slender to fairly slender (2–2.5mm). Medium to fairly long (17–25mm). Protrudes beyond base.
Cavity Quite deep and cone shaped becoming quite narrow towards the centre with the stalk sunk well within. There can be some grey-brown russet.
Eye Medium size. Closed or slightly open. Sepals tapering, erect-convergent with some tips reflexed. Very downy.
Basin Medium width. Medium depth. Usually five definite ribs and some slight puckering. Usually russet free occasional tiny specks of russet.
Tube Cone-shaped.
Stamens Median.
Core line Basal, clasping.
Core Median. Abaxile or axile.
Cells Mostly ovate. Sometimes rather elliptical.
Seeds Acuminate or acute. Numerous. Straight.
Flesh White tinged green. Firm. Coarse-textured. Juicy.
Aroma Nil.
Flowers Pollination group 3.
Leaves Medium size. Rather long acute. Serrate or bluntly serrate. Medium thick. Flat or very slightly undulating. Upward-folding. Mid yellowish-green. Slightly downward-hanging. Downy.

GROWING APPLES

H.A. Baker

Origin

The cultivated apple, *Malus pumila domestica*, is not a true species but a hybrid of complex parentage. Its taxonomy has been obscured by a process of hybridization, selection and rejection by man and nature over thousands of years to such an extent that no one today can say what its exact ancestry is. Its evolvement is a continuing process. Modern geneticists with the objectives of breeding desirable qualities, such as hardiness, heavier cropping, pest and disease resistance, etc. into the apple are still introducing other *Malus* species into its make-up. But, the large fruited apple as we know it today is thought to have descended from the crab apple *Malus pumila*.

It is surmised that the apple was first domesticated in the region just south of the Caucasus, as large fruited crab apple hybrids as well as distinct species still exist in primeval forests in that area. It spread from the Caspian Sea across Europe to the Atlantic in pre-historic times. Evidence of such fruits have been found in Neolithic dwellings in Central Europe. Because of its complex parentage, and the wealth of genetic material within its make-up, the apple shows more variability in its progeny than any other major fruit. It does not grow true from seed: every seedling is different. No other fruit has such variability in flavour, texture, shape, colour, or season. There is a cultivar to suit every taste.

Thus if you plant a pip, say from a Cox's Orange Pippin, and it germinates, that seedling will be unique, unnamed and yours to do with as you will. One word of caution to dampen any feelings of optimism about filling the world with wonderful new apples; the chances of raising a seedling better than the many thousands of existing named cultivars is exceedingly remote – a chance in a thousand in fact. Remember, those we have today have come to the top by the hard Darwinian process of rejection and selection carried out by our forefathers and it is still going on.

This is not to say that no good new apple occurs; it does, but not very often and more often than not from professional geneticists based at research institutes. Having said this, two of the most important apples in the United Kingdom at the present time, Cox's Orange Pippin and Bramley's Seedling, were raised by amateurs well over a hundred years ago.

Climate

The apple is a deciduous tree growing in the cool, temperate regions of the world; though, because it has been domesticated by man for such a long time, and because of its great adaptability, there are a few cultivars which will grow and yield quite successfully in the warm, temperate and even in the sub-tropical regions. In the UK and the US it can be grown virtually anywhere given reasonable conditions. Areas subjected to strong winds, exposed coastal regions for example, are unsuitable unless of course ample wind breaks are provided. Strong winds can damage and distort growth, reduce the soil temperature, blow blossom and fruit to the ground and inhibit the movement of those essential pollinating insects. Similarly regions of high altitude present difficulties. The greater the altitude, the cooler the weather and the shorter the growing season. Fruits from such places are smaller, greener, less sweet than those from warmer areas.

In general terms the best apples are grown at altitudes of less than 600 feet, ideally below 400 feet, where the summer temperatures are high and the rainfall low. Commercial apple growing tends to be concentrated in East Anglia, Kent and the southern counties as far west as the Vale of Evesham and to the borders of Devon.

Professional growers are aiming for perfection. Nevertheless, as already mentioned, the apple is a most adaptable fruit of great diversity and there are few areas in this country where some cultivar or other cannot be grown. Apples which are especially hardy, and therefore suited to the cooler regions, have been noted in the text.

In similar vein, areas of high rainfall have problems because, unfortunately, the apple can be attacked by fungal diseases associated with wet conditions, in particular apple scab, *Venturia inequalis*. This is the disease which causes black scab-like lesions on the skin of the fruit and, in bad attacks, cracking, which in turn allow the ingress of fruit rotting diseases. Scab can also attack the wood and likewise lead to the wood rotting disease apple canker, *Nectria galligena*. Apple scab is more prevalent in the wet areas but is ubiquitous and can occur anywhere whenever warm wet conditions occur. There are a few apples which are resistant, or partially so, to scab and some less prone to canker, and this is noted in their descriptions. Additionally modern science has provided us with fungicidal chemicals capable of preventing or eradicating scab. It is, therefore, possible to grow scab prone apples, such as Cox's Orange Pippin, in areas where otherwise they could not be contemplated. It is appreciated that not every gardener is keen on using chemicals and there is still the problem of how to spray a large tree. Growers reluctant to use chemicals should not be deterred but consider growing the scab resistant cultivars and cooking apples, or be prepared to accept the blemished fruit. After all, the skin can be peeled if necessary.

Soil and Soil Drainage

The ideal soil is a slightly acid, about pH6.7, well drained, medium loam eighteen inches or more in depth. However, the apple is tolerant of a wide range. Good drainage is the operative term, as waterlogged soil leads to all kinds of problems such as root death, poor growth, low yield, apple canker and possibly the complete loss of the tree. Badly drained ground should be avoided and where this is not possible some kind of drainage system must be installed. Shallow soils over chalk also are unsuitable because of the problems of lime induced chlorosis and the poor water holding capacity of such land. Gardeners on shallow soils should be prepared to dig out the chalk to a depth of not less than fifteen inches by a two foot radius at each planting site and replace it with a good medium to heavy loam. Light soils are acceptable but their moisture retentiveness must be improved by the generous use of organics and with irrigation whenever necessary.

Frost

One of the greatest hazards to successful apple growing is frost at blossom time. Severe frosts at this critical period can destroy the whole or part of the potential crop. When the buds are dormant they are safe, but once they are opened they become increasingly vulnerable. The answer is, in the choice of site, to avoid a frost pocket. Cold air is denser than warm and will therefore gravitate to the lowest point, pushing the warm air upwards as it does so. Areas where cold air collects are called frost pockets. They may be natural, valley bottoms for example, or they may be man-made, a hedge or solid wall impeding the escape of cold air. Remember, if there is any danger of a frost pocket being created, when planting or constructing a hedge or fence, make provision for cold air to flow away. For instance, leave a twelve inch gap at the bottom of a hedge or erect a slatted fence rather than a solid one.

Avoiding a frost pocket is easier said than done:

the garden is where the house is. However, all is not lost in such a situation as there are ways and means of mitigating or avoiding the damage, and much depends on the ingenuity of the grower. Covering the trees whenever frost is forecast is the obvious answer, but if the tree is large this is obviously impracticable. Therefore, whenever new plantings are contemplated, it would be wise to grow trees in restricted form, and on dwarfing rootstocks, so that they will always be small enough to cover if necessary. In areas which are particularly prone, choose cultivars which flower later than normal and therefore stand a better chance of avoiding those late spring frosts.

Choosing the Site

Select a sunny sheltered position for any new planting. Sunshine and warmth are necessary to ripen the wood, promote the development of fruit buds and give size, colour, flavour and sugars to the apple. The importance of shelter has already been emphasized. Suffice to say, it is essential if a good fruit set and subsequently a satisfactory crop of high quality fruits are to be obtained.

Choice of Cultivar and the Requirements of Cross Pollination

There are so many lovely apples, as will be seen when looking at the ensuing descriptions and pictures, that one is almost spoilt for choice. Remember, when selecting apples for the garden, that no cultivar is truly self-fertile. It will not set a good crop with its own pollen. This means that an apple should not be planted singly but have a partner as pollinator, another but different cultivar which flowers at the same time. Pollinating insects, such as bees, will perform the essential service of cross pollination at blossom time. If there is insufficient room for more than one tree, then plant a family tree, that is, a tree created by the nurseryman with more than one cultivar grafted upon the rootstock. It might consist of two, sometimes up to four, cultivars which are specially chosen both so that cross pollination is achieved and to give a selection of dessert fruits or perhaps a mixture of dessert and culinary. It is not a wise policy to rely on a neighbour's tree to pollinate yours, but if there is no other recourse it can sometimes happen that pollination by such means is perfectly adequate. There is a limit to how far a bee will travel in its pollen and nectar-seeking forays before it returns to the hive, therefore the tree must not be too far away. As a rough guide, cross pollination will be effective up to sixty feet becoming gradually less effective thereafter.

For pollination purposes, apples are grouped together according to their time of flowering. The

Dwarf Bush

Bush

Half Standard

Standard

earliest flowers are in group 1 and the latest in group 7. (See pollination table, page 142.) In selecting apples for cross pollination, ideally choose those within the same group; an alternative is to choose from those groups immediately adjacent as there will be sufficient overlap. Forecasting the time of flowering cannot be an exact science as there will be variations from year to year, according to the district and the climate.

There is another important point to consider. Most apples are Diploid in their genetic construction, meaning that they can produce fertile pollen in the normal course of reproduction. However, a few apple cultivars are Triploid, which means they have one extra set of chromosomes. Such apples are perfectly fertile on the female side, that is they bear fruit, but they are very poor producers of pollen. Bramley's Seedling is an example. Where these are planted, select two pollinators to pollinate the triploid and each other. Triploids are designated with a т on the pollination table. Incidentally, triploids are usually very vigorous and are best grafted on to one of the dwarfing rootstocks.

There are a few apples which will set a reasonable crop with their own pollen. No real work has been done on the subject, so there is no comprehensive list, but from observation and experience, it can be said that the following come within this category:

Beauty of Bath	Worcester Pearmain
Emneth Early	Chiver's Delight
Keswick Codlin	Ellison's Orange
Charles Ross	Lord Derby
Greensleeves	Newton Wonder
James Grieve	Crawley Beauty
Sunset	

Choice of Tree Form

There are two basic categories of apple trees as far as form is concerned. The first are those grown in the open and pruned in the winter; the typical orchard trees in effect. These are the bush, half standard, full standard and spindle bush. The latter is a form now widely used especially by professional growers. The second category are those grown in restricted form and pruned in the summer, such as the cordon, dwarf pyramid, espalier and, more rarely, the fan. These are the forms best suited to the small suburban garden where space is limited, though a small orchard is still a popular feature in the larger garden. Each form is covered in detail below.

Bush, Half Standard and Full Standard

The bush is the most widely grown of the orchard forms. It is an open centred goblet-shaped tree on a short trunk of about two and a half feet. The dwarf bush is a smaller version with a trunk of about one and a half feet, and is grafted on to one of the dwarfing rootstocks.

The half and full standards are merely taller versions of the bush. The basic shape is the same, except that the trunks are longer, and the heads bigger. The half standard has a trunk of about four and a half feet, and the standard, six feet or more. The standards are grafted on to vigorous rootstocks and are in most instances unsuitable for the modern garden because of their size. Large trees can crop very heavily, perhaps too heavily, and their vigorous habit can create problems with pruning, picking and spraying. Nevertheless, there can be a place for the large tree as a feature or to provide shade.

Spindlebush

Sometimes called the centre leader tree. Its branches start at about fifteen to eighteen inches from the ground. Those at the base are the longest, those at the top are the shortest, so that it is cone-shaped. The spindlebush is usually grafted on to one of the dwarfing rootstocks.

The centre leader tree has certain advantages over the open centre bush. The wide angle which the branches make with the central stem are strong and less liable to break under the weight of fruit. Horizontal growth is more fruitful and less vigorous than upright and much of this type of growth is achieved by tying down the young laterals where horizontal growth does not occur naturally. The disadvantage of the spindlebush is that a tall stake is necessary to support it, and the string or weights tying down the laterals make the tree look rather ugly in the garden. Additionally, a strict regime of renewal pruning is necessary to keep it productive and under control. It is only for the enthusiast, because if the top growth is allowed to become too dominant it can be very difficult to correct without major tree surgery. All this means is that it is a form only suitable for the skilled fruit grower.

The Restricted Forms

Cordon

This is the most widely planted of the restricted forms and is ideal for the small garden. Being closely spaced, i.e. two and a half to three feet apart, means a goodly number of cultivars can be planted in a relatively small area. Moreover, as apples are not self fertile the essential requirement of cross pollination is easily covered.

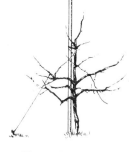

Young Spindlebush

Established Spindlebush

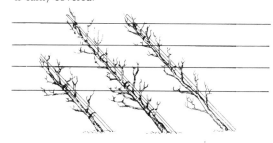

The oblique cordon planted at an angle of forty-five degrees, rather than vertical cordons, is to be preferred. Top growth on a vertical cordon can grow away too strongly at the expense of the lower fruit spurs. Planting at an angle is a happy compromise between growth and fruitfulness and it also gives you extra length of cropping stem relative to height. Cordons are ideal for growing against the wall or fence. Out in the open a post and wire fence is necessary to support them, with the top wire at about six feet.

Espalier

The espalier consists of horizontal arms or tiers, more or less opposite each other, arising from a central vertical stem. It needs more lateral room than the cordon because of its outstretched arms. It can be a low or high form if necessary, depending upon the number of tiers. A two tier espalier for example, can be kept to a height of between three and four feet, whereas a four tier espalier needs a height of about six and a half feet. Espaliers make attractive boundary markers between one part of the garden and another, around the vegetable plot, for example, or as a hedge between the vegetables and the ornamentals. It is a decorative form, the outstretched arms look handsome in the spring, covered in blossom, and pleasing in the autumn, covered with fruit.

Dwarf Pyramid

This is similar in shape to the spindlebush except that it is not as tall, being kept down to a height of no more than six feet by summer pruning. The dwarf pyramid is always grafted on to one of the very dwarfing rootstocks. It is closely spaced and again ideal for the small garden, but it needs to be planted out in the open and not against fences and walls as with the cordon and espalier. The dwarf pyramid needs support, which can either consist of an individual stake to each tree or by the use of two horizontal wires at eighteen and thirty-six inches to which the trees are tied.

Dwarf Pyramid

Fan

The fan is an unusual form for the apple, being mainly reserved for stone fruits, which is a pity because it is an attractive shape. Where the garden has walls or fences of seven feet high or more, the fan is an alternative form to the espalier or cordon.

Because the apple does not grow true from seed the only way of perpetuating and multiplying a desired cultivar is by vegetative means. It is an interesting thought that of the many thousands of Cox's Orange Pippin trees in existence today they all stem from one single apple seedling raised by Mr Richard Cox at Colnbrook Lawn, Slough in 1825.

To take a cutting from a selected tree might seem the obvious way of propagation but most apple cultivars do not root readily by this method. For this reason, and others, the apple is propagated by grafting. Scion wood from the chosen cultivar is budded or grafted on to a compatible rootstock. The rootstock, more than any other factor, governs the eventual size of the mature tree, and it is important to make the correct choice, so that the tree or trees are right for the space available and the tree form in which they are to be grown. Most commercial orchards are closely planted and farmed intensively and, sadly perhaps, most modern gardens are small, therefore the rootstocks mainly used by nurserymen today are dwarfing or semi-dwarfing. Incidentally, these stocks have the added advantage of making the tree precocious, so they bear fruit quickly. The principal rootstocks in order of vigour are:

M.27 Extremely dwarfing. Suitable for the dwarf centre leader trees and cordons. Requires a fertile soil and trees need support throughout their lives. It is excellent for vigorous cultivars but unsuitable for weak ones. It is not a stock which will tolerate neglect. Very precocious.

M.9 Very dwarfing. Suitable for dwarf bushes, pyramids, spindle bushes and cordons. It requires a good soil and the trees need staking throughout their lives.

M.26 Dwarfing. A good all-round rootstock suitable for all forms, including the small espalier. Ideal for average soil conditions.

MM.106 Semi-dwarfing. The stock most widely used by nurserymen because it is suitable for most forms and soils, including the lighter ones. Trees out in the open on this stock need staking for the first four or five years.

MM.111 & M.2 Vigorous. Suitable for half and full standards. They could be used for the restricted forms such as the espalier and cordon but only on poor soils otherwise their vigour would become a problem.

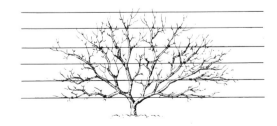

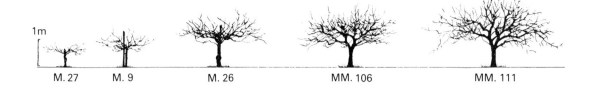

M. 27   M. 9   M. 26   MM. 106   MM. 111

Spacing Relative to Tree Form

Spacings given below are not precise but intended as a guide only, because tree size will vary to some extent depending upon the cultivar, environment and the growing conditions.

Orchard Trees

Rootstock	Dwarf bush	Bush	Half & Full Standard	Spindlebush
M.27	4' to 6'	–	–	4' to 6'
M.9	8' to 10'	–	–	6' to 8'
M.26	–	10' to 12'	–	8' to 10'
MM.106	–	12' to 15'	–	8' to 12'
MM.111 & M.2	–	–	15' to 20'	–

Restricted Forms

Rootstock	Dwarf Pyramid	Cordon	Espalier	Fan
M.27	4' to 6'	2½'	–	–
M.9	6'	2½'	–	–
M.26	6'	2½'	10'	10'
MM.106	6' to 7'	2½' to 3'	12' to 15'	12' to 15'
MM.111 & M.2	–	–	15' to 18'	15' to 20'

Planting in the Open

Soil Preparation and Planting

The best time to plant is when the trees are dormant: ideally in the autumn immediately after leaf fall in November, or in the early spring in March before the trees move into growth and after the worst of the winter is over. It is not a good policy to plant in the middle of winter when the soil is wet, cold or frozen. In the meantime, heel in the trees in a sheltered spot. It is most important that the roots do not dry out or are subjected to frost. If the ground is frozen solid and it is not possible to heel the trees in, put them in a garden shed or somewhere cool. Keep the roots covered so that they do not dry out but unwrap the aerial parts so they are not forced prematurely into growth. Prepare the ground well. Remove all perennial weeds, by the use of chemical weed killers if desired. Where grass land is involved double digging is necessary, or, for large scale apple growing,

Trees heeled in

deep ploughing. Turn the turves upside down. It would be wise to kill the grass first, especially if couch is present, using a weed killer. Where closely spaced trees are to be planted, turn the whole plot over but for single trees it is sufficient to prepare the land over a radius of about two feet at each planting site.

Medium to heavy fertile land will not need much bulky organics in the planting hole and it will suffice to fork in a bucketful of peat, a handful of bonemeal and four ounces of a balanced fertilizer. Light soils will need more generous treatment, say two bucketfulls of peat plus the other materials mentioned. The bulky organics are essential to improve the moisture retentiveness of the soil and its structure.

As already mentioned, the ideal time to plant is in the dormant season, using bare rooted trees dug up with a substantial root system, the kind supplied by a reputable fruit tree nurseryman. Containerized plants, whilst they have the advantage that they can be planted at any time even in the growing season, unfortunately often, have the root system drastically reduced in order to fit into the container.

Plant the tree to the same depth as it was in the nursery. The soil mark on the stem of the tree will be an indicator. It is important that the union between rootstock and scion is well above soil level, not less than four inches, to avoid the risk of scion rooting. Many nurserymen graft at about twelve inches so there is no danger. Remember, whilst the tree is out of the ground the root system must not be allowed to dry. If necessary the tree should be heeled in in a sheltered spot until wanted and, throughout the planting operations, keep the roots covered either with soil or with a piece of damp hessian.

All apple trees, no matter what their form, should be supported for the first years at least. Trees on the very dwarfing rootstocks and those grown in restricted form require permanent support. Cordons, espaliers and fans, for example, are tied to canes affixed to wires.

For trees out in the open, measure the size of the root system and dig out a hole slightly larger and deeper. Drive the stake in first and then plant the tree to the stake. Place the tree three inches away from the tree stake to allow for the eventual expansion of the tree trunk. Ensure that there are no branches chafing against the stake on that side. With bush trees it is usual to drive the stake in so that the head is clear of any branches on that side. Replace the soil, and firm as planting proceeds so that it

Maiden Whip

makes good contact with the roots. Finally, secure the tree to the stake with a figure eight tie. One can buy proprietary tree ties which provide a cushion between the stake and the tree so there is no danger of rubbing. The last task is to mulch with bulky organic material to help soil moisture retention. Mulch over a radius of fifteen to eighteen inches to a depth of about two inches but keep it just clear of the tree stem to avoid any danger of collar rot.

Pruning

It is beyond the intention of this book to go into great detail about pruning. Suffice to say that the restricted forms such as cordon, espalier, dwarf pyramid and fan are pruned in the summer. Its purpose is to inhibit their growth, necessary because they are grown in a confined way or in limited space, hence the term restricted. Summer pruning checks growth. The removal of leaves at the height of summer cuts down the supply of carbohydrates to the roots and therefore reduces the overall energy of the plant. It has other advantages too. It allows light and air on to the wood and on the fruit, helping to promote fruit buds for the next year and to improve the colour of the existing fruits. Winter pruning is applied to trees grown in the open, typically the bush, standard and spindlebush. Hard winter pruning has the effect of stimulating strong growth in the next growing season, hence is only usually applied to young trees whilst the framework is being formed. Established trees which are cropping well are pruned lightly.

Feathered Maiden

It is not a wise policy to become hide-bound to one pruning system but apply whatever technique is necessary as circumstances dictate. There are occasions when summer pruning can be applied to trees normally pruned in the winter: the over vigorous, unfruitful bush or standard for example. Similarly, restricted forms can be winter pruned to restore a neglected cordon or espalier back to its original form or to reduce the length of over-long spur systems.

2 year-old tree

Spring and Summer Tasks

Feeding

It is necessary, on a regular basis, to replenish the main elements extracted from the ground by the tree for its growth and development and the creation of its crop. These are nitrogen (N), phosphorus (P) and potassium (K). In general terms nitrogen is necessary for growth, phosphorus (phosphates) for root development and fruit quality, and potassium (potash) for hardiness, fruit bud formation, fruit flavour and colour. There are also occasions when the minor elements, magnesium and calcium are needed. The trace elements are utilized in minute quantities and are therefore rarely needed except under certain special circumstances.

3–4 year-old tree

The most efficient way of supplying the trees' main nutrient requirements is to use the artificial fertilizers because they are concentrated, not bulky, and therefore easy to handle. They have their disadvantages, however, in that they do nothing for the soil structure, moisture retentiveness and very little for the soil fauna and flora, i.e. the life of the soil itself.

Nitrogen is usually applied in the form of sulphate of ammonia (21% N) on neutral to alkaline soils, and ammonium nitrate/lime (nitro-chalk) (21% N) on acidic soils. Potassium is applied as sulphate of potash (48% K_2O), and phosphorus as superphosphates (18–19% P_2O_5) or triple phosphates (47% P_2O_5). Nitrogen and potash are applied annually and the phosphates triennially because the latter is used up much more slowly.

Apply the artificial fertilizers as a top dressing over the rooting area which is roughly equivalent to the spread of the tree and slightly beyond as follows: sulphate of potash in late January at three quarters of an ounce per square yard. Superphosphates every third year in January at two ounces per square yard. Triple superphosphates at three quarters of an ounce per square yard also in late January. Sulphate of ammonia or ammonium nitrate/lime (nitro-chalk) at one ounce per square yard.

Dessert apples growing in grass and cookers are given twice this rate of nitrogen. In the case of dessert apples, to off-set the competition of grass for the nitrogen, and in the case of cookers to improve the size of the culinary fruits.

Turning now to the minor element magnesium. This is a constituent of the chlorophyll molecule necessary for the manufacture of sugars by the leaf. The trouble is, it is easily leached out of the leaves by rain, therefore in wet summers and in areas of high rainfall expect a magnesium deficiency and act beforehand. Apply magnesium sulphate at twice the rate advocated for potassium, i.e. one and a half ounces per square yard in early April. If, despite this, magnesium deficiency manifests itself, and it will in very wet summers, supplement by applying two or three foliar applications of magnesium sulphate (Epsom Salts) at fortnightly intervals. This is usually throughout the months of June and early July. Artificial fertilizers supply the necessary nutrients efficiently, but bulky organics also have their place in the feeding programme. Whilst they are not high in nutrient content compared with artificials, they are invaluable in other important respects. They maintain the life of the soil – the soil's fauna and flora – which in turn create and maintain the soil structure and its retentiveness. They also break down the various organic materials into a form which can be assimilated by the roots.

Closely spaced plants which are competing with each other, such as the restricted forms, will certainly benefit from a mulch of well rotted manure, compost, peat or mushroom compost. Spread along

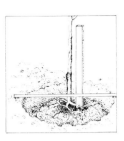

Planting young tree

each side of the row or around the plant to a depth of about two or three inches but always keep just clear of the stem of the tree so there is no danger of rotting diseases being encouraged. The mulch should be topped up annually in the spring.

The young tree in the orchard should also be mulched for the first four years. Thereafter the orchard can be grassed down so long as a clean area of about eighteen inches radius is maintained around the trunk of each tree. Remember, grass competes with the tree for nitrogen and extra nitrogen will be necessary, certainly for the cooking apples to give a good size.

Watering

Water in times of drought, especially the newly planted and young trees not yet fully established. Water is essential so that the trees grow well and form a strong framework. Just as importantly, irrigate established trees carrying a heavy crop. Trees under stress due to lack of moisture are liable to suffer excessive fruit drop and go into a biennial pattern of bearing, producing a crop every other year. The relevant period for irrigation is the summer months of June, July and August. Apply about two inches (nine gallons per square yard) of water over the rooting area about once every seven to ten days starting in June and finishing in late August. Obviously stop when the rain restores the balance. To lessen the risk of fungal disease, apply the water over the ground rather than the foliage.

Thinning

Thinning is necessary in the event of a heavy fruit set to avoid too many small fruits and to lessen the risk of the onset of biennial bearing to which certain cultivars are prone. Remember, there will be a natural shedding of fruitlets and this takes place in late June and the first half of July. It is called the June drop. The main thinning should be done after this event but a little can be carried out before by removing any malformed and diseased fruits. Thin dessert apples to about four to six inches apart, and culinary fruits, where a larger size is wanted, to six to nine inches. Obviously, remove the worst and leave the best. The King fruit, i.e. the one in the centre of the truss, is usually the largest and can be left, but check that it is not malformed at the stalk end. Weak trees should be thinned more drastically than the strong ones. Vigorous trees with a good show of leaves can be allowed to carry more fruit than the guidelines just given.

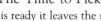

Thinning

The Time to Pick

When the fruit is ready it leaves the spur easily. Lift the apple in the palm of the hand and give it a slight twist when it should part without undue effort. Avoid any finger pressure as this will bruise and spoil the appearance of the fruits. For the same reason, at all times treat the apples gently. Bruised fruits will not keep.

Other indications that the harvest is nigh are that the fruits take on brighter colours, and the pips turn from white to brown. Windfalls on the ground are another indication that they are ready, gales aside of course. Do not pick all the fruit at once, but practise colour picking, which means picking over a number of times, starting with those exposed to the most sun, leaving the inside fruits to the last so that they may eventually colour up.

Storage

Early apples do not keep, they should be eaten straight off the tree or within a few days of picking. July and August apples come within this category. The essential conditions required for good keeping are coolness, darkness, high humidity and ventilation. Not too much ventilation or the fruits will shrivel, nor too little or they will be destroyed by lack of oxygen. Do not keep mid-season apples with the lates because the volatile substances (that lovely smell they give off) will hasten the ripening of the lates. Store each cultivar separately as much as possible.

The fruit store should be able to maintain a cool, even temperature, ideally at about 37°F (2.7°C). Commercially this is done by refrigeration. On an amateur basis, the best one can usually achieve is about 42° to 47°F (5° to 7°C). A well built garden shed situated in the shade is fine. A garage comes second best, so long as frost does not penetrate the store and it is secure against mice. Do not keep apples in the loft as quite high temperatures can build up in this area when the sun is shining.

The construction of the container in which the apples are kept is just as important. It must allow air movement through the fruits and over the top. Wooden apple boxes with slatted sides and base and corner posts are ideal and so are wooden Dutch tomato trays. Store only sound fruit and inspect them regularly to remove rots. Remember all apple cultivars have their season: an October apple is at its best in this month; before then it is immature and after, it is over the top. A cultivar with a season January to April, Sturmer Pippin for example, is ready to eat within this period and not before. Another very good way of keeping apples is in a polythene bag. This material maintains high humidity and so prevents the fruit from shrivelling too quickly. However, the apple must be allowed to breathe. The skin of the bag should be perforated with a hole the diameter of a pencil for every pound of fruit, and the top of the bag folded over rather than sealed. Use clear polythene so that the apples can be seen and any rots removed if necessary. The required conditions of coolness, darkness and ventilation still apply.

Aphids

Woolly aphid

Winter moth caterpillar

Apple sawfly

Fruit tree
Red Spider Mite

Pests and Disease Control

Unfortunately apples can be inflicted by a vast range of pests and diseases, not forgetting our feathered friends, the Bullfinch in the winter, and the Blackbirds, Thrushes and Starlings in the summer. Space permits mention only of the most important. Incidentally, never spray insecticides at blossom time because of the risk to bees and other pollinating insects. To spray or not to spray, that is the question. For those gardeners that want perfect fruits, destruction of the pathogens by chemical means is the usual answer at the present time. Scientists are aware of the public's growing reaction to the use of such chemicals and are working towards non-chemical means of control: the use of predators and parasites are typical examples. For the non-perfectionist, indeed for any good gardener, the other way of dealing with the problem is to practise garden hygiene. Keep things tidy and clean. Keep the trees properly fed and pruned. Remove or burn the prunings, rots, and other debris which might harbour trouble. Having said this, unsprayed trees will not yield perfect fruits and the gardener will have to accept that this is in the natural order of things.

Insects   In order of their appearance.

It is in spring that the over-wintering eggs of aphids and caterpillars hatch out. Once hatched they migrate to the opening flower and leaf buds.

Aphids   are extremely debilitating creatures because they suck the sap out of the plant. They have the capability of multiplying rapidly and can very quickly infest the whole tree. Signs of their attack are severe leaf curling, stunted and distorted growth and dwarfed malformed fruits. Spray with a systemic insecticide.

Woolly Aphid   is another aphid, very different in its appearance and mode of attack. It does not over-winter in egg form but as a complete insect. It is rather an immobile aphid which usually lives in the crevices of the older wood, in wounds, cracks and under the bark. The Woolly Aphid is covered with a woolly-like substance, presumably as a form of camouflage, hence the name. Other indications of its attack are large, gall-like swellings on the wood where it lives. Spray with a systemic insecticide, apply a winter wash, or paint isolated patches with insecticide.

Winter Moth Caterpillars   are looper caterpillars which eat the spring foliage and the developing flowers. Damage can be serious. Spray with a contact insecticide.

Apple Sawfly   the larvae – white maggots – either tunnel into the fruitlets which then drop, or graze a line along the surface, usually around the stalk end, causing a ribbon-like scar. Spray with an insecticide at petal fall.

Fruit Tree Red Spider Mite   a sap-sucking insect which lives on the underside of the leaves and which

thrives in a hot summer. It is very small and a hand lens is the best way of seeing it. Red Spider Mite is a very difficult insect to control because of its capability of quickly developing resistance to chemicals. It is best left to natural predators.

Codlin Moth   the most important of the insect pests. The larvae is a caterpillar which eats into the centre of the apple causing a most unsavoury mess. The moth lays her eggs in mid to late June and the time to spray is in mid June and again three weeks later, using a contact insecticide.

Diseases   the most important are scab, mildew and canker. There are a few cultivars which have some degree of resistance to one or other of these diseases and this is mentioned in the cultivar descriptions.

Apple Scab   a fungus which causes black scab-like lesions on the skin of the fruit and, in severe cases, cracking and malformation. It also attacks the leaves and sometimes the wood. On the leaves it manifests itself as brown or olive green spots resulting in early leaf fall. On wood it shows as small blister-like pimples on the young shoots and can later lead to the wood-rotting disease apple canker. Scab is worst in a wet summer and in areas of high rainfall. Spray with a protectant/erradicant fungicide on a regular fortnightly basis from bud burst to mid July.

Apple Mildew – Powdery Mildew   a fungal disease mainly of the apple foliage, though it can dull and russet the skin of the fruit. It shows as a powdery or mealy coating of the leaves and shoots as they move into growth in the spring. Mealy rosettes of leaves with little or no growth is another sign. If left uncontrolled, mildew will persist and multiply throughout the tree during the growing season and is a most debilitating problem. Cut off and burn every infected shoot as they become obvious in the early Spring. Adopt a regular spray programme from bud burst to mid July on either a weekly or fortnightly basis, using a fungicide advocated for mildew.

Apple Canker   another serious disease of apples to which certain cultivars are prone, for example James Grieve, Spartan and Cox's Orange Pippin. It is more prevalent on trees growing in badly drained land and in wet areas. Canker is a wood rotting disease, showing first as sunken oyster-shaped areas, which if left uncontrolled will eventually girdle and kill the infected branch and obviously, if on the main trunk, can result in the death of the tree. Cut out the diseased wood and paint the wound with a fungicidal tree paint. Girdled, dead or dying branches must be removed and burnt. An untreated canker is a reservoir of infection.

Bitter Pit   a disease that appears on the fruits. Some cultivars are very prone: for example Newton Wonder, Bramley's Seedling and Egremont Russet. It produces slightly sunken pits on the surface of the skin and small brown areas in the flesh immediately beneath the pits and scattered throughout the apple. It is thought that the onset of this disease is caused

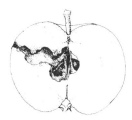

Codlin Moth

Apple scab

Apple mildew

Apple canker

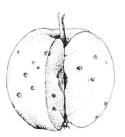

Bitter pit

by the lack of water at a critical time and a calcium deficiency. It can be partially avoided by mulching well and watering in dry periods.

Recommended Cultivars

The list below indicates the apple cultivars, both dessert and culinary, which are generally considered to be of the highest quality flavour. In most cases, this has been established through reputation over a great number of years but it also includes some more recent and promising introductions.

Dessert Apples

Adam's Pearmain	Jupiter
Ashmead's Kernel	Kidd's Orange Red
Blenheim Orange	Laxton's Epicure
Cornish Gilliflower	Laxton's Fortune
Cox's Orange Pippin	Lord Hindlip
D'Arcy Spice	Lord Lambourne
Discovery	Merton Charm
Egremont Russet	Mother
Fiesta	Orlean's Reinette
Gala	Pixie
George Cave	Ribston Pippin
Greensleeves	St Edmund's Pippin
Holstein	Spartan
Irish Peach	Sunset
James Grieve	Suntan
Jonagold	William Crump

Culinary Apples

Annie Elizabeth	Golden Noble
Blenheim Orange	Monarch
Bountiful	Newton Wonder
Bramley's Seedling	Norfolk Beauty
Encore	Rev. W. Wilks
George Neal	Warner's King

Pollination Table

т = Triploid
в = Biennial or irregular in flowering

Group 1

Gravenstein (т)	Stark's Earliest

Group 2

Adam's Pearmain (в)	Lord Lambourne
Beauty of Bath	Margil
Bismark (в)	McIntosh Red
Devonshire Quarrenden (в)	Merton Charm
Egremont Russet	Norfolk Beauty
George Cave	Owen Thomas
George Neal	Rev. W. Wilks (в)
Idared	Ribston Pippin (т)
Irish Peach	St. Edmund's Pippin
Keswick Codlin (в)	Warner's King (т)
Laxton's Early Crimson	

Group 3

Allington Pippin (в)	Kidd's Orange Red
Arthur Turner	Lane's Prince Albert
Belle de Boskoop (т)	Laxton's Epicure
Belle de Pontoise (в)	Laxton's Fortune (в)
Blenheim Orange (т)(в)	Lord Grosvenor
Bountiful	Lord Hindlip
Bramley's Seedling (т)	Malling Kent
Brownlees Russet	Mère de Ménage
Charles Ross	Merton Knave
Cox's Orange Pippin	Merton Worcester
Crispin (т)(в)	Miller's Seedling (в)
Discovery	Nonpareil
Duchess's Favourite	Peasgood Nonsuch
Emneth Early (в)	Queen
Emperor Alexander	Red Delicious
Fiesta	Rival (в)
Granny Smith	Rosemary Russet
Greensleeves	Spartan
Grenadier	Stirling Castle
Hambledon Deux Ans	Sturmer Pippin
Holstein (т)	Sunset
James Grieve	Tom Putt
John Standish	Tydeman's Early Worcester
Jonathan	Wagener (в)
Jupiter (т)	Wealthy
Katy	Worcester Pearmain

Group 4

Annie Elizabeth	Golden Noble
Ashmead's Kernel	Harvey
Autumn Pearmain	Herring's Pippin
Barnack Beauty	Howgate Wonder
Chiver's Delight	Jonagold (т)
Claygate Pearmain	King's Acre Pippin
Cornish Aromatic	Lady Sudeley
Cornish Gilliflower	Lady Henniker
Cox's Pomona	Laxton's Superb (в)
D'Arcy Spice	Lord Burghley
Duke of Devonshire	Lord Derby
Dumelow's Seedling	Monarch (в)
Ellison's Orange	Orlean's Reinette
Encore	Pixie
Gala	Tower of Glamis
Gladstone (в)	Tydeman's Late Orange
Golden Delicious	Winston

Group 5

Gascoyne's Scarlet (т)	Norfolk Royal
King of the Pippins (в)	Royal Jubilee
Merton Beauty	Suntan (т)
Mother	William Crump
Newton Wonder (в)	

Group 6

Bess Pool	Edward VII
Court Pendu Plat	

Group 7

Crawley Beauty

CLASSIFICATION TABLE

Key

Numbers indicate months in which the apple should be eaten

c = Culinary	‖ = Fairly prominently ribbed	d = Dual purpose	‡ = Indefinitely ribbed
		# = Boldly ribbed	+ = No ribs or slight trace

FLAT ⌒	ROUND ◯	CONICAL ⌂	OBLONG AND OVAL ⬭ ⬮
GROUP 1 GREEN-SMOOTH SKINNED – ACID			
Stirling Castle 9/12 +	Grenadier 8/10 ‖	Emneth Early 7/8 #	
Bramley 11/3 ‖		Lord Grosvenor 8/10 #	
	Warner's King 9/2 ‖ *(between Round and Conical)*		
	Edward VII 12/4 +	Lord Derby 10/12 #	
	Tower of Glamis 11/2 # *(between Round and Conical)*		
GROUP 2 GREEN-SMOOTH SKINNED – SWEET			
	Granny Smith 1/4 ‡		Greensleeves 9/11 +
GROUP 3 WHITISH CREAM or YELLOW SKIN – SWEET or ACID			
	Norfolk Beauty 10/12 ‖	Keswick Codlin 9/10 # c	Greensleeves 9/11 +
	Golden Noble 10/12 +	Rev W Wilks 9/10 ‡ c	Royal Jubilee 10/12 # c
		Harvey 9/1 ‡ c	Golden Delicious 11/2 ‡
		Arthur Turner 9/11 ‡ c	Crispin 12/2 # d
GROUP 4 GREEN or YELLOW SKIN – FLUSHED and/or STRIPED – ACID			
	George Neal 8/10 ‡ *(between Flat and Round)*		
	Tom Putt 9/11 # *(between Flat and Round)*		
Queen 9/12 ‡	Monarch 11/1 +	Howgate Wonder 10/3 ‡	
	Newton Wonder 11/3 + *(between Flat and Round)*		
	Dumelow's Seedling 11/3 + *(between Flat and Round)*		
Belle de Pontoise 11/3 ‖			
Bramley 11/3 ‖	Lane's Prince Albert 12/3 ‡		Encore 12/4 ‡
	Crawley Beauty 12/3 + *(between Flat and Round)*		Annie Elizabeth 12/6 +
GROUP 5 GREEN or YELLOW SKIN – FLUSHED and/or STRIPED – SWEET and SUB-ACID			
Owen Thomas 8/9 #	Irish Peach 8/9 ‡	Lady Sudeley 8/9 ‡	
	Beauty of Bath 8 + *(between Flat and Round)*	Miller's Seedling 8/9 ‡	
	Laxton's Fortune 9/10 ‡	Laxton's Epicure 8/9 +	
	Merton Charm 9/10 +	James Grieve 9/10 ‡	
	Bountiful 9/11 ‡ c	Emperor Alexander 9/11 + c	Gravenstein 9/12 #
	Lord Lambourne 9/11 +		
	Peasgood Nonsuch 9/12 + c		
	Rival 10/12 ‡ *(between Flat and Round)*	Cox's Pomona 10/12 # c *(between Round and Conical)*	
		Charles Ross 10/12 + *(between Round and Conical)*	
	Chiver's Delight 11/1 ‡	Sturmer Pippin 1/4 ‡	
Wagener 12/4 ‖	Jonagold 11/2 + *(between Round and Conical)*		

FLAT ◌	ROUND ◌	CONICAL ◌	OBLONG AND OVAL ◌ ◌

GROUP 6   OVER HALF FLUSHED BRIGHT RED

FLAT	ROUND	CONICAL	OBLONG AND OVAL
Discovery 8/9 +	Stark's Earliest 7/8 +	Laxton's Early Crimson 7/8 +	
	Duchess's Favourite 8/9 +	Gladstone 7/8 ⧺	
Devonshire Quarrenden 8/9 ⧺	Tydeman's Early Worcester 8/9 ‡	Katy 9/10 +	
		Worcester Pearmain 9/10 +	
	Merton Knave 9 +	Herring's Pippin 9/11 ⧺	
	Merton Worcester 9/10 +		
	Wealthy 9/12 ‡	Norfolk Royal 9/12 ‡	
	Merton Beauty 9/10 +		
Fiesta 10/1 +		Spartan 10/2 ‡	Gala 10/1 ‡
	McIntosh Red 10/12 ‡	Mother 10/11 ‡	
	Gascoyne's Scarlet 10/1 ‡	Jupiter 11/1 ‡	
Mère de Ménage 11/2 ⧺		Bismark 11/2 ⧺	Jonathan 11/2 ⧺
Idared 11/4 ‡		Malling Kent 11/2 ‡	Red Delicious 12/3 ⧺
	John Standish 12/2 +	William Crump 12/2 ‡	

GROUP 7   REINETTES

FLAT	ROUND	CONICAL	OBLONG AND OVAL
	Ellison's Orange 9/10 +	Autumn Pearmain 9/11 ‡	King of the Pippins 10/12 +
	Ribston Pippin 10/12 ⧺	Allington Pippin 10/12 +	
	Sunset 10/12 +		
Suntan 11/1 +	Margil 10/1 ⧺		
Blenheim Orange 11/1 +	Cox's Orange Pippin 10/1 +		
Orlean's Reinette 11/1 +		Kidd's Orange Red 11/1 ‡	Lady Henniker 11/1 ⧺ d
		Laxton's Superb 11/1 ‡	
		Holstein 11/1 +	
		Rosemary Russet 11/3 ‡	Cornish Gilliflower 11/3 ⧺
		Adam's Pearmain 11/3 +	
		Hambledon Deux Ans 11/4 ‡	
	Bess Pool 12/2 +		Claygate Pearmain 12/2 ‡
Pixie 12/3 +		Cornish Aromatic 12/3 ⧺	Barnack Beauty (Oval) 12/3 +
		Lord Hindlip 12/3 ⧺	
		King's Acre Pippin 12/3 +	
		Belle de Boskoop 12/4 ‡	
Court Pendu Plat 12/4 +		Tydeman's Late Orange 12/4 +	
	Lord Burghley 1/4 ‡	Winston 12/4 +	

GROUP 8   RUSSETS

FLAT	ROUND	CONICAL	OBLONG AND OVAL
	St Edmund's Pippin 9/10 +		D'Arcy Spice 12/4 ⧺
Egremont Russet 10/12 +			
Ashmead's Kernel 12/2 ‡			
Brownlees' Russet 12/3 ‡			
	Nonpareil 12/3 +		
	Duke of Devonshire 1/3 +